Telecommunications: A Systems Approach

Telecommunications: A Systems Approach

by
G. SMOL B.Sc., Ph.D., C.Eng., M.I.E.E.
M. P. R. HAMER M.A.
M. T. HILLS B.Sc., Ph.D., C.Eng., M.I.E.E.

LONDON GEORGE ALLEN & UNWIN LTD
Ruskin House   Museum Street

First published in 1976

ISBN 0 04 621022 9   HARDBACK
ISBN 0 04 621023 7   PAPERBACK

Printed in England
at The Lavenham Press Limited, Lavenham, Suffolk.

Preface

The aim of this book is to introduce basic ideas on telecommunication systems to students at undergraduate level by means of two extended case studies on telephone and television systems. The ideas covered are relevant to the operation of switched systems in general, not just telephones, and to the television systems of both Europe and America. Wherever possible agreed international practice, as embodied in the recommendations of the International Telecommunication Union, has been described.

The choice of case studies allows us to deal with a wide range of techniques appropriate to switched and non-switched, narrow and broad band, analogue and digital systems, and provides a practical context within which the characteristics and purposes of each technique may be appreciated. A more conventional approach, and one most suitable for a reference text, would be to present each technique as part of a list ordered under general headings, such as data transmission or statistical methods. We believe that our approach is preferable when the aim is to introduce techniques, because it highlights their essential features and the reasons why they are needed.

The book was written as part of an Open University telecommunications course. It constitutes a self-contained element of the course providing an introduction to telecommunication systems, and it assumes a knowledge of what may be described as elementary telecommunication principles: a.c. circuits; amplification and feedback; amplitude, frequency and pulse code modulation; basic transmission line theory; and elementary noise concepts. The book can therefore provide the basis for the systems component of a telecommunications course at undergraduate or at a similar level.

This book owes a great deal to the inspiration and encouragement of Professor J. J. Sparkes. We wish to express our thanks to him and to members of the Open University Telecommunication Systems Course Team, particularly Dr D. I. Crecraft, for reading all the drafts we have produced and contributing numerous helpful suggestions. We are also indebted to Mr J. R. Pollard and Mr R. Beaufoy of Plessey Telecommunications Limited and to Professor J. A. Turner of the University of Essex for their advice and their detailed comments on the final draft. We would like to thank members of the British Post Office who have contributed directly or indirectly to some of the material in the book. Finally we would like to thank Mrs Christine Martindale and Miss Marlene Rose for typing the many drafts of this book with great speed and accuracy.

Contents

Chapter 1

Introduction and overview

The aim of this book is to introduce and explain some of the basic ideas and techniques used in telecommunication systems and to indicate how they may be applied. The material it contains has been selected in the belief that this aim can best be achieved by concentrating on a fairly detailed description of specific examples; and that the alternative, a nominally comprehensive but very superficial coverage of the whole field of telecommunication systems, provides a much less satisfactory foundation for an understanding of the subject.

This point of view has led to the selection of two systems: telephone and television. However, a few simple examples of other systems are described in the first chapter in order to introduce some of the ideas which will be taken up in greater detail later on, and also to provide a very brief survey of the subject before concentrating on the two chosen systems.

1.1 A SIMPLE SYSTEM

Telecommunication involves communicating at a distance using some form of equipment. The distance can be as short as a few metres. Two people holding a conversation over an intercom in two different rooms of the same house are using a telecommunication system. They would not be involved in telecommunication if they were just holding a normal conversation in the same room or even shouting from one room to another. The equipment involved in telecommunication systems does not have to be electrical; it could, for instance, be acoustic, as in the case of a voice tube. However, this book only deals with systems using electrical equipment because these are, by far, the most common.

It is convenient to introduce basic ideas through the description of a simple system. Compared with many systems an intercom is indeed simple, but it is a two-way system—both users can either talk or listen. A baby alarm, shown in Figure 1.1, is somewhat similar to an intercom, but it is even simpler because it is a one-way system. It comprises a microphone, placed in the baby's room, and a loudspeaker in a room occupied by the baby's parents. Sounds produced by the baby are converted into electrical signals by the microphone, these signals are conveyed down the cable, amplified by the amplifier and converted back into sound, that is pressure waves in air, by the loudspeaker. The microphone and loudspeaker are the **input** and **output transducers.**

The two transducers and the elements between them constitute a telecommunication system. The human beings, the baby and its parents, are not included in the system. They are referred to as **users.** They provide system

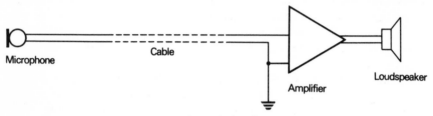

Figure 1.1 A baby alarm

inputs and utilise the system outputs. We have chosen, as a matter of convenience, to draw the system boundaries between the transducers and the users, although the characteristics of users play a very important part in telecommunication systems, as will be stressed throughout this book.

The cable is the **transmission channel.** The amplifier can be located at either end of it. The amplifier can be thought of as forming part of a terminal which also comprises one of the transducers. Thus in Figure 1.1, the amplifier forms part of the receiver **terminal,** and the transmitter terminal is simply a microphone. The system elements of the baby alarm are shown in Figure 1.2.

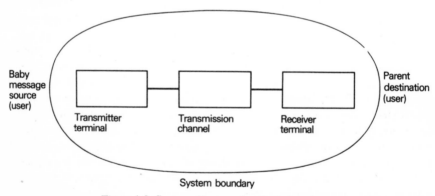

Figure 1.2 System diagram for the baby alarm

It will often be convenient to break down much more complex systems into elements like these which can be thought of as 'black boxes'. They are shown linked by single lines because the diagram makes no attempt to indicate the details of their connections. In dealing with telecommunication systems we consider each black box as a separate element interacting with the other elements of the system. We are concerned with the functions and specifications of the black boxes, with the way in which their outputs relate to their inputs, and how they affect the overall system; but we are not usually concerned with their minute internal details. We may, however, find it necessary to treat a black box as a subsystem which can be broken down into further black boxes.

We can look at the baby alarm amplifier as a black box. We will not normally be concerned with the details of the amplifier circuit, but we will want to know such things as its location within the system, its bandwidth, its gain, the distortion and noise it introduces and the power it can deliver to the loudspeaker.

It is worth considering a few of the factors which lead to the choice of some of these amplifier parameters, because they illustrate the part that user requirements and properties play in the choice of system parameters.

Location of the amplifier

One might not want to put a mains operated amplifier in a baby's room because of possible risks of fire or electric shock. The same safety considerations would not apply with an amplifier powered by, say, a 9 V dry battery. One might choose to put such an amplifier in the baby's room so that only microphone signals are amplified and not cable noise due, for instance, to electrical interference caused by domestic equipment. Signal-to-noise ratio and safety are therefore some of the factors involved in deciding whether to put the amplifier in the baby's or the parent's room.

Dynamic range

Let us assume that the parents not only want to hear the baby's cries but that they also want to hear its breathing when it is sleeping normally, because this provides a check that all is well and that the system is working. The ratio of the acoustic powers produced by the baby crying at its loudest and breathing quietly might well be of the order of 70 dB. This is the **dynamic range** of the message source. Such a dynamic range would stretch the capabilities of the most expensive high-fidelity systems but, fortunately, it is not necessary to achieve this range. A faithful reproduction of the breathing and of the cries is not really required. All that is needed is for the parent to be able to monitor the breathing and recognise the crying. The amplifier needs to have enough gain at low input levels, that is, **small-signal gain,** to make the breathing audible. At high inputs the amplifier may behave non-linearly, giving reduced gain and causing distortion of the signal, but this does not matter, provided only that cries should be recognisable as such. This requirement can be met by an amplifier which would be described as rudimentary by high-fidelity standards.

Bandwidth

The information conveyed by the system is an electrical signal which we will call the **message signal.** In the present case it corresponds to crying or breathing. The signal is corrupted by various forms of noise, such as electrical interference picked up in the cable, or thermal noise generated in various parts of the system. The noise can produce an audible output throughout the whole frequency range of the loudspeaker. The largest frequency components of the message signals are contained in a band of about 1 kHz to 3 kHz. Thus, if instead of extending over the whole range of frequencies which can be picked up by the microphone and reproduced by the loudspeaker, amplification is effectively limited to this 1-3 kHz band, both crying and breathing will be reproduced quite adequately with a significant improvement in signal-to-noise ratio at the output of the system. It is therefore possible to reduce the effect of noise at frequencies where there is little message information by suitably limiting the amplifier bandwidth. This makes it possible to achieve a sufficiently low background noise level to make breathing audible despite the use of a relatively simple amplifier and of a cable that is not elaborately screened.

The required small-signal gain and output power of the amplifier

The required small-signal gain of the amplifier will depend on (a) the sensitivity of the loudspeaker, which can be specified as the electrical input power it needs to produce an adequately audible output for the minimum required signal level (corresponding to the baby's breathing); (b) the sensitivity of the microphone, which can be similarly specified as its electrical power output for the minimum required signal level; and (c) the loss due to the cable in the frequency band of interest.

For instance if the minimum power output of the microphone is, say, 0.02 mW, the power corresponding to minimum signal level which needs to be delivered to the loudspeaker is 20 mW and the cable loss is 3 dB (i.e. half the power is lost in the cable); then the required amplifier gain is

$$10 \log \frac{20}{0 \cdot 02} + 3 = 33 \text{ dB}$$

The output power of the amplifier, corresponding to the minimum signal level, depends on the location of the amplifier. Using the above figures, the minimum required output power is 20 mW if the amplifier is next to the loudspeaker. If the amplifier is next to the microphone, the output power of the amplifier must be 40 mW because half of it is lost in the cable. Note that the location of the amplifier does not affect the required gain because the cable introduces the same amount of loss in both cases.

Besides providing examples of some of the basic elements of a telecommunication system, this discussion of a very simple system illustrates how the characteristics of the users serve in determining the required characteristics of the system elements. Although the loudspeaker output is far from being a faithful replica of the sounds produced by the baby, we can still recognise this output as breathing or crying. The choice of amplifier bandwidth and permissible non-linearity was based on an estimate of the frequency spectrum of the message source and of the relevant features of human perception: in this case our ability to recognise particular sounds despite band limitation and distortion. By using this knowledge it is possible to choose the simplest and, therefore usually, the cheapest design which will satisfy the system requirements.

User perception characteristics are of general importance in all telecommunication systems. The common purpose of these systems is to convey information from the message source to the user. We are not concerned here with the precise nature of this information, either from a philosophical point of view or in terms of the abstract ideas of information theory. Our concern is with the user's reactions. Does he find the system satisfactory? On a telephone, for example, can he hold an intelligible conversation? Ultimately our estimate of user reactions can only be obtained from **user tests** of systems or parts of systems.

It is often impossible to satisfy everybody with a single set of parameters. For instance, Figure 1.3 shows the results of tests made by the British Post Office on the most suitable receiver power level for a telephone system. It can be seen that if the level of reception is, for instance, 10 dB below the reference level used in the experiment, about 26 per cent of users say it is too

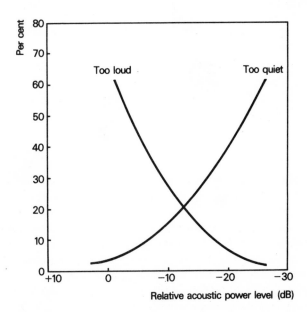

Figure 1.3 Results of listening tests to establish loudness preferences. The curves show the percentage of subjects finding a given level too loud or too quiet (after Richards, D. L. 1973, *Telecommunication by Speech*, London: Butterworth)

loud and about 16 per cent of users say it is too quiet. It can also be seen from the figure that no level satisfies everybody. It is therefore necessary to find characteristics which satisfy as many normal users as possible. This means making a sufficiently large number of tests to obtain statistically valid results.

User tests are concerned with the user's overall perception. They do not involve assumptions about the operation of the human ear, eye or brain as separate entities, only the overall response of the user to a particular form of physical stimulus. It is the characteristics of this type of overall response which will be referred to as **perception data** in this book. Perception data will be seen to provide the starting point in the selection of many system parameters. For instance, the minimum acceptable bandwidth and signal-to-noise ratio in a telecommunication system are determined on the basis of human perception properties as deduced from user tests. In general, the complexity and cost of a system increase with both bandwidth and signal-to-noise ratio, and it is wasteful to provide more bandwidth, or a higher signal-to-noise ratio than is strictly necessary on the basis of perception data. Other quantities, for instance many of the parameters of monochrome and of colour television systems, are chosen on this basis. This will be discussed in Chapters 5 and 6.

1.2 SOME EXAMPLES OF TELECOMMUNICATION SYSTEMS

We will now consider the uses of telecommunication which are listed in Table 1.1. Our purpose is to introduce some of the terms, concepts and techniques treated in this book.

The first example is of a local, own-exchange, telephone call, that is a call between two users whose telephones are connected to the same local exchange. The exchange is connected to telephone terminals in private homes, offices, factories, and so on, by local lines. The exchange, terminals and lines can be thought of as a system. We can ignore the fact that the exchange is connected to other exchanges by trunk lines because we have chosen to look only at an own-exchange call.

Telephone calls can be set up between any pair of terminals and several calls can be set up simultaneously between separate pairs of terminals. Each call is **two-way,** that is both users can act simultaneously as message source and destination. They are linked by a two-way transmission channel over which they can both listen and talk. There is one telephone terminal at each end of the channel so that it is a **one-one call.** However, by using special loudspeaker telephones it is possible to have more than one person involved in the conversation at each end. Such a call would still be described as one-one on the basis of the number of terminals involved. **Many-one** and **many-many** calls are possible with extension telephones and conference calls.

The setting up of a call involves connecting two terminals by the operation

Table 1.1 *Examples of telecommunication*

Type of communication	Message source	Transmitter terminal	Transmission channel	Receiver terminal	Destination
Own-exchange telephone call	Speech	Telephone	Local lines and exchange	Telephone	Human listener
Transatlantic telephone call	Speech	Telephone	Local and trunk lines, local, trunk and international exchanges, communication satellite or submarine cable	Telephone	Human listener
Public broadcast radio	Speech and music	Studio, r.f. transmitter and links between the two	Radio waves	Radio receiver	Human listener
Telegraph	Human operator	Teletypewriter	Lines (wire pairs) and exchanges	Teletypewriter	Human addressee
Data link to a computer	Human operator or paper tape	Teletypewriter and modem	200 baud data link	Modem and computer input/output devices	Computer

of a number of switches in the exchange, either automatically or by an operator. This way of setting up calls is characteristic of **switched systems.** The baby alarm we considered originally is an example of a one-way non-switched system.

There are many types of switched systems, and their use is widespread in telecommunications. They will not be discussed in any detail in the present chapter because they form a major part of the subject matter of Chapters 2, 3 and 4.

The second example in Table 1.1, a transatlantic telephone call, also involves a switched telephone system, only now we are dealing with an international, or even world-wide, system. The particular call we are considering uses similar terminals to the first example. The difference lies in the transmission channel which now involves a large number of elements which will be referred to as the **transmission links** forming part of the channel. They include local lines, trunk lines, several types of exchange and a transmission link spanning the Atlantic. This consists of a submarine cable, or of a radio subsystem in which message signals from either terminal are transmitted to a satellite where they are received and retransmitted towards their destination. The satellite orbit is chosen so that, using suitable antennae, radio beams can be propagated simultaneously between the satellite and both continents.

Both satellite and submarine cable transmission channels are designed so that they can be used for hundreds or even thousands of simultaneous telephone calls. This involves the technique of multiplexing, without which the cost of transatlantic telephone calls would be prohibitive. **Multiplexing** consists in combining a number of message signals so that they can be transmitted over the same channel.

Frequency division multiplexing (f.d.m.) is used on transatlantic links. A telephone message signal occupies a bandwidth of about 3 kHz, but submarine cables and satellites can transmit signals whose bandwidth is very much greater than this. Both types of channel can be made to carry several thousand telephone calls by taking the signal for each call and frequency shifting it. This is done by modulating the signal onto a carrier. A different carrier frequency is chosen for each call so that, when multiplexed, it occupies a different band from all the others. The frequency shifting is achieved by modulation, in fact by single sideband (suppressed carrier) amplitude modulation.

The signals are multiplexed at the transmitter end and demultiplexed (returned to their original frequency bands by a filtering and demodulation process) at the receiver end.

The third telecommunication example in Table 1.1 is a radio broadcast system. It may be divided into elements in such a way that the transmission channel is conceived as being between the transmitter antenna and the receiver antenna. This makes the transmitter terminal into a complex subsystem involving studios, control rooms, possibly telephone lines to a distant radio transmitter and the radio transmitter itself. This is an arbitrary division, for instance the telephone lines and radio transmitter could be included as part of the transmission channel.

The radio transmitter serves two main purposes:

(1) To convert the message signal into a form suitable for propagation as a radio wave. This involves modulation to produce a radio frequency (r.f.) signal. Some public broadcast radio systems use amplitude modulation (a.m.), others use frequency modulation (f.m.).

(2) To amplify the r.f. signal so that, when propagated from the transmitter antenna, it will provide adequate signal strength at the receiver antennae throughout the region served by the transmitter.

A national radio system usually has several transmitters broadcasting the same programme material. Each transmitter covers different parts of the country. Broadcast systems are one-way and one-many.

The last two examples in Table 1.1 will be considered in greater detail because, unlike telephones and radio, they are not normally used in the home and may, therefore, be less familiar to you. The fourth example, a telegraph system, has terminals consisting of teletypewriters which are often referred to by the registered trade name Teleprinter. They are fitted with keyboards, somewhat like typewriters, and a paper roll on which alphanumeric characters, that is letters, numbers and punctuation signs, may be typed. Each teletypewriter consists of a transmitter and a receiver section. When a key is pressed the transmitter section produces a series of direct current pulses which can have one of two magnitudes. Each pulse can therefore be thought of as a binary digit, or bit, with one amplitude corresponding to a binary 1 and the other to a binary zero. The terms mark and space are often used for 1 and 0 pulses respectively. They date back to the times when the output of telegraph receivers consisted of a series of marks and spaces produced on a paper tape by the up and down motion of a pen operated electromagnetically from the transmitter terminal. We will use 1 and 0 in this book.

The output of a teletypewriter is an example of a **digital signal.** Digital signals are made up from a finite number of discrete levels, e.g. two levels corresponding to 0 and 1. They differ in this respect from **analogue signals,** such as the output of a microphone which can vary continuously over its dynamic range.

The pulses produced by the transmitter section of one terminal are sent along wires and used to operate the receiver section of the second terminal which behaves effectively as an electric typewriter operated from the transmitter end.

The speed at which teletypewriters operate is often stated in **bauds,** which are units of pulse rate. If the duration of the shortest pulse which the system can send is t seconds, then $1/t$ pulses can be transmitted per second. This is the pulse rate in bauds. The definition of the baud (pulse per second) includes time in the same way as the hertz (cycle per second). Teletypewriter speeds range from 45·5 to 150 bauds. Since the pulses are limited to two discrete levels, the bit rate of a teletypewriter signal is equal to the speed in bauds. A common speed is 50 bauds. A standard instrument of this speed uses five pulses per character and two additional pulses, the start and stop pulses, used to indicate the beginning and end of each character. The five character pulses and the start pulse are equal in length, but the stop pulse is 1·42 times longer than the others. This is to allow stop pulses to be identified automatically. Thus a character takes up 7·42 pulse periods so that the teletypewriter can send 60 × 50/7·42 characters per minute, or 60 × 50/(7·42 × 6) ≃ 67 words per minute, where one word is defined as six characters (five letters and a space).

The telex (*tele*graph *ex*change) system uses teletypewriters and sends signals over switched networks involving exchanges. The figure keys 0-9, or dials associated with the teletypewriters can be used to set up calls automatically between telex terminals, both on national and international networks. In order for this to be possible, the terminals and transmission channels must be **compatible.**

Compatibility requires that the teletypewriters must operate at standard pulse rates and use the same binary codes for their characters. This question of compatibility applies to all telecommunication systems which have widespread use. Compatibility between systems of different countries is made possible by the work of the International Telecommunication Union (ITU) which has been in existence for over 100 years. It is now part of the United Nations and works through a number of committees consisting of representatives of all the member nations. These committees include the CCITT (Comité Consultatif International de Téléphonie et de Télégraphie) dealing with telephony and telegraphy, and the CCIR (Comité Consultatif International de Radiocommunication) dealing with radio. These committees publish their recommendations on technical standards, operating procedures, accounting, etc., in a series of books roughly every four years. Each series has a different colour and is often referred to by that colour. The recommendations are not mandatory but they are adopted by most administrations, not only because they facilitate inter-working between systems, but because they lead to standard specifications for the manufacture of telecommunications equipment and thus ensure wider markets for this equipment. However, this is not always so. For instance, confusion can arise in the case of teletypewriter operation because the terminology used in North American telecommunications and in most of the computer industry throughout the world does not accord with the CCITT definitions.

Consider our system consisting of two teletypewriters linked by a transmission channel. It may be possible to send different messages in both directions simultaneously; it may be possible to send messages in both directions, but only one way at a time; or it may only be possible to send messages in one direction.

The terms used in North America and in the computer industry for these possibilities are:

Full duplex   If simultaneous two-way operation is possible.

Half duplex   If operation in either direction is possible, but only one way at a time.

Simplex   If only one-way operation is possible.

The CCITT terms are mainly concerned with the nature of the transmission channel (circuit). They are:

Duplex circuit   If simultaneous two-way operation is possible.

Half duplex circuit   If the channel allows duplex operation, but the terminals are so constructed that transmission is only possible in one direction at a time.

Simplex circuit   If the channel allows transmission in either direction, but only one way at a time.

Channel   If the channel only allows transmission in one direction.

Teletypewriter systems often use various forms of multiplexing. Frequency division multiplexing is used with both amplitude and frequency modulation of the carriers. If a carrier is amplitude modulated with teletypewriter pulses which are effectively square waves, a very wide spectrum of sideband frequencies is produced, but the bandwidth of the modulated signal is suitably reduced by passing it through a filter. However, this has the effect of delaying the build up of carrier amplitude and it can be shown (Reference 3 pp. 117-20) that the build up time, t, is approximately equal to the reciprocal of the bandwidth, B, of the filter.

A signal must have duration of at least t seconds to be correctly received, so that the maximum number of pulses which can be transmitted per second is B. Thus, a 50 baud signal requires a bandwidth of at least 50 Hz. But this is only an approximate value, obtained for an ideal system, and a higher value is used to cater for the use of non-ideal filters, the variation of carrier frequencies and the effects of noise. The CCITT recommended value for 50 baud systems is 120 Hz per channel, and teletypewriter channels are normally multiplexed in groups of 12, 18 or, more recently, 24 onto one telephone speech channel, using carriers at 120 Hz intervals.

When frequency modulation is used for f.d.m., carriers 120 Hz apart are shifted up and down in frequency by 30 Hz for 1 and 0 pulses. This type of modulation, in which the carrier frequency is shifted by discrete steps, is called **frequency shift keying** (f.s.k.).

Another form of multiplexing, **time division multiplexing** (t.d.m.) can also be used. It can be described in terms of the simple scheme of Figure 1.4 which represents how the outputs of four teletypewriters can be combined into a single signal.

Each terminal operates at 50 bauds and therefore produces pulses of 20 ms duration. These are stored in a register in the multiplexer. A circuit is used to sample the stored pulses, taking one pulse from each terminal in turn. The output of this circuit is a pulse of the same type (1 or 0) as the original, but with only a quarter of its duration. Thus, as shown in Figure 1.4, the four separate sets of input pulses become interleaved into one set of shorter pulses which can be sent along a single channel.

The reverse process, demultiplexing the train of pulses, is carried out at the other end of the channel by electronic logic circuits which separate out the pulses and use them to generate four sets of 20 ms pulses which are transmitted to four receiving terminals.

The pulse patterns used for t.d.m. are normally more complicated than Figure 1.4. Extra bits, known as framing bits, may be added to identify the terminals and the first pulse of each character. Also, the sampling may be carried out over complete characters instead of individual pulses.

The fifth example from Table 1.1 is a data system consisting of a computer accessed from a remote terminal. The terminal is a teletypewriter fitted with equipment to produce and read punched paper tape. The computer instructions and data can be prepared off-line; that is the teletypewriter can be disconnected from the line and used to prepare a data tape which can be checked for mistakes before sending. Once the tape has been prepared, the terminal can be reconnected to the line and the data transmitted using the tape reader. During transmission the teletypewriter at the terminal is often used to monitor the outgoing data. The computer output is returned along the same line and is either printed out directly or punched onto paper tape for subsequent off-line printing.

A pulse rate of 110 bauds is commonly used on data links, with a total of 11 bits (binary pulses) per character. This includes one bit for the start and two for the stop signals, seven bits for the different characters, and one parity bit, which allows for a simple error check. The value of the parity bit is chosen to ensure that the seven character digits together with the parity bit only contain an even number of 1 bits. Thus an error, due, say, to noise on the line, which changes a 1 into a 0, or vice-versa, will be detected because the received signal, which has an odd number of 1 bits, no longer corresponds to any of the alphanumeric characters used in the system.

A suitable transmission channel for the system is a CCITT standard data channel designed for full duplex operation at pulse rates up to 200 bauds. The signals for the two directions are multiplexed onto the channel by frequency shift keying. The calling terminal uses 980 Hz for 1 pulses and 1180 Hz for 0 pulses. The called terminal uses 1650 Hz for 1 pulses and

1850 Hz for 0 pulses. A modulator is used at the terminal to convert the out-going pulses into the appropriate audio frequency signal and a demodulator is used to convert the incoming audio frequency into interrupted d.c. pulses to operate the teletypewriter or tape punch. The *modulator* and *dem*odulator are usually housed in a single unit called a **modem.**

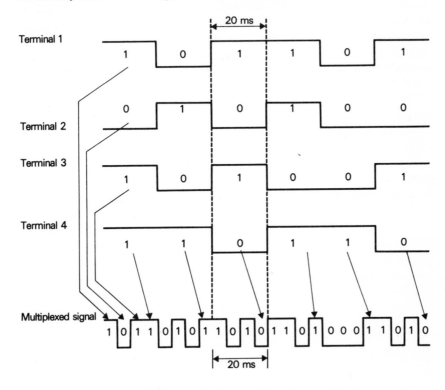

Figure 1.4 Time division multiplexing of four binary digital signals

1.3 TELECOMMUNICATION SERVICES

Telecommunication systems are used to provide a wide variety of services: public, private, military, and so on. They are far too numerous to list in detail, but will be roughly categorised in this section in terms of the transducers they use at their transmitter and receiver terminals, and the type of message information they convey.

Where appropriate, the required transmission channel bandwidth, or bit rate in the case of digital systems, will be mentioned. The bandwidth and bit rates of transmission channels are limited by many factors. For instance radio channels can be used only over frequency bands that are not already occupied by other transmissions in the same geographical area. Also,

25

telephone cables can only usefully carry signals of limited bandwidth because their attenuation increases with increasing frequency. In many systems the cost of bandwidth is a major constraint: this is why bandwidth is mentioned in the following list. There are of course many other parameters, such as minimum acceptable signal-to-noise ratio, error rate, required dynamic range, reliability and useful operating life time, but stating figures for these would require too many qualifications at this stage. Bandwidth considerations are so often overriding that they are worth mentioning, even in an introduction. The figures given are approximate. They refer to the 3 dB bandwidth, but the sharpness of the band edges depends on the nature of the system.

(a) *Teletypewriter links*

A typical example of a teletypewriter used for telegraph and telex services was given in the previous section. It operated at 50 bauds which was equivalent to 67 words per minute, each word consisting of five characters and a space. The required bandwidth was 120 Hz. Teletypewriters are made with rates of up to 150 bauds. They can therefore be used with the CCITT standard transmission channel which can accommodate pulse rates up to 200 bauds.

(b) *Telephone links*

Telephone systems represent the most extensive and varied application of telecommunication techniques using every type of transmission channel, from a single wire with earth return, to communication satellites. An audio frequency bandwidth of about 3 kHz is generally considered adequate. The CCITT standard voice band is 300-3 400 Hz.

Besides national, international, military and private telephone systems, there are many special systems using mainly radio in connection with mobile terminals, for instance aircraft and ship communications, police, ambulance and taxi services, and radio amateurs.

(c) *High-quality voice and music transmissions*

Although a bandwidth of 3 kHz is adequate for recognising speech, we can perceive sound at higher frequencies (up to 20 kHz in the case of young people with normal hearing) and a bandwidth of about 15 kHz is considered necessary for high quality music transmissions. This is provided in some f.m. public radio broadcast systems.

(d) *Static visual material*

Photographs, drawings, printed or written pages, weather maps, finger prints, etc., can provide the message source for a variety of systems which produce some form of photocopy sheet at the message destination. This is called **facsimile.** The picture to be transmitted is mounted on a rotating cylinder which is scanned by a photocell mounted on a lead screw linked to the cylinder. An electrical analogue signal is produced which varies as light and dark parts of the picture are scanned. At the receiver the analogue signal is amplified and made to operate a light source of variable intensity which scans a photosensitive sheet mounted on a cylinder rotating in synchronism with the transmitter cylinder. When the scan is completed the photosensitive sheet is developed and used to produce a copy of the message source. Various bandwidths are used. Larger bandwidths allow more material to be sent in a given time. For instance, a page of a newspaper takes about 25 minutes if a bandwidth of 40 kHz is used. Doubling the bandwidth approximately halves the time required for transmitting the same material.

(e) *Graphic terminals*
A variety of devices has been developed to convey diagrams over telephone links, particularly for remote teaching and telephone conference purposes. The transmitter terminals consist of a pad on which the diagrams are drawn with a special pen whose position is detected by appropriate sensors. These generate electrical signals which are transmitted to the receiver terminal where they are used to reproduce the diagram on the face of a cathode ray tube or on a sheet of paper by means of a chart plotter. The 3·1 kHz bandwidth of the telephone link limits the rate at which the picture can be produced. (The maximum drawing speed for one device is of the order of 10 cm s^{-1} on a 30 $\times$ 21 cm pad.) The picture therefore builds up relatively slowly. The picture information is held in an electronic store at the receiver. This enables the picture to be displayed on the screen for as long as it is required.

(f) *Vision phones*
When two people hold a face-to-face conversation they exchange a good deal of information by visual means through gestures and changes of facial expression. This information, which is lost in a normal telephone call, can be provided by means of an additional television link. However, the service proves to be very expensive because of the large bandwidth required compared with simple telephone calls. The Picturephone system developed by the Bell Telephone Laboratories of America uses a bandwidth of 1 MHz and provides a picture 12·5 by 14 cm made up of 267 horizontal lines.

(g) *Closed circuit television*
This involves the transmission of television signals over cables and can be used for a variety of purposes, such as remote surveillance of road traffic conditions or of equipment in hazardous environments such as nuclear reactors. Bandwidths ranging from 1 to 5 MHz have been used.

(h) *Broadcast television*
This includes broadcast radio transmissions and cable television. Cable television is used in regions of poor radio reception. A single antenna, located

at a suitably high point feeds an amplifier or receiver whose output is distributed over cables to all the users in the district. Cable television is also used to provide local programmes which could not be broadcast by radio without interfering with reception in other districts.

The video signal of the UK 625 lines service has a bandwidth of 5·5 MHz but, as is discussed in Chapter 5, Section 5.10, there is a spacing of 8 MHz between programme channels. This is to allow for the sound signals and for the extra bandwidth required for the form of vestigial sideband modulation used to modulate the video signal onto an r.f. carrier.

(i) *Data*

This can take a wide variety of forms. Direct digital links between computers typically require up to 1 Mbits s^{-1}. The CCITT recommended rates for data networks are 200, 600 and 1 200 bauds. The recommended frequencies for frequency shift keying are 1 300 Hz (1 pulse) and 1 700 Hz (0 pulse) for the 600 baud system; 1 300 Hz (1 pulse) and 2 100 Hz (0 pulse) for the 1 200 baud system. The figures for the 200 baud system were given at the end of Section 1.2. The rapid growth of a variety of computer-like machines which can store and process digital information is leading to an ever increasing range of applications for data networks. These include as typical examples

Telemetry: measurement at a distance in which the outputs of all manner of transducers such as accelerometers, pressure gauges, altimeters, thermometers are encoded and transmitted to a remote terminal.
Remote control.
Data collection: sales and inventory control, bank accounting using a central processor.
Data dissemination: stock exchange prices, weather information.
Direct interaction with a distant computer: air-line booking, on-line computer access from a remote terminal.

1.4 TRANSMISSION MEDIA

The media used for transmission channels can be classified into two broad categories:

(i) bounded (for example along electrical conductors)
(ii) radio.

Twisted pairs, coaxial cables, waveguides and optical fibres are bounded media.

1.4.1 Bounded media: cables and waveguides

Wire pairs are widely used in telephone and telex networks. They are sometimes carried overhead on poles, but this is no longer practicable in cities and is unsightly in the country, so buried cables consisting of a large number of twisted wire pairs (up to 4000) are more commonly used. The cable pairs are twisted in order to reduce the stray electric and magnetic fields which can couple to neighbouring pairs causing **crosstalk** between calls.

Twisted wire pairs are used for single links between telephone terminals and local exchanges. Their use over broad band multiplexed channels is limited, because both the attenuation and crosstalk increase with increasing frequency. Twisted wire-pair cables have been used up to 500 kHz with appropriate amplifiers, called **repeaters,** at regular intervals, but coaxial cables are used for higher frequencies.

A coaxial cable consists of a pair of concentric cylindrical conductors held in position by insulators. Several coaxial cables are usually bunched together inside a common protective sleeve. The outer conductor of each cable acts as an electrical shield; but the effectiveness of this shield decreases at low frequencies and below about 60 kHz crosstalk on bunches of coaxial conductors becomes excessive. They are therefore used for multiplexed systems operating at frequencies above 60 kHz. Coaxial cables have been used up to 61·6 MHz to carry 10800 telephone channels. They are used on long distance links with evenly spaced repeaters. The repeater spacing depends on bandwidth. For instance, a CCITT standard cable having a 2·9 mm inner and a 9·5 mm outer diameter requires a repeater spacing of 9 km for an upper frequency of 4·1 MHz and 1·5 km for an upper frequency of 61·6 MHz. A submarine cable consists of a single coaxial cable with repeaters sealed in at regular intervals. Cables providing from 48 to 3600 f.d.m. telephone channels have been used.

Radio waves can propagate along hollow conducting pipes, known as **waveguides,** providing that the cross sectional dimensions of the pipe are of the order of, or greater than, the wavelength of the radio waves. Practical waveguides can be made with cross-sectional dimensions of the order of centimetres or millimetres. Radio waves of this range of wavelengths are called **microwaves.** Waveguides are used in microwave repeaters using frequencies in the range 2 GHz to 11 GHz, corresponding to wavelengths of 15 cm to 2·7 cm, and systems have been proposed for frequencies up to 80 GHz. Waveguides do not transmit signals below a certain frequency, the cut-off frequency, which depends on the waveguide dimensions. Even so they can be used over bandwidths in excess of 20 per cent of their operating frequencies. Thus a 10 GHz system may have a bandwidth of more than 2000 MHz, which can accommodate more than 500000 telephone channels. Telephone channels are each allowed 4 kHz in f.d.m. systems even though they only occupy 3·1 kHz. This is in order to avoid having to use very sharp cut-off filters when demultiplexing, because such filters are expensive. Only 2·85 kHz (200 to 3050 Hz) is allowed in the case of long distance submarine cables. This is because the slight reduction in sound quality due to the

29

decreased signal bandwidth and the cost of the sharp cut-off filters at the two ends of the cable are more than outweighed by the advantage of having extra channels over an extremely expensive transmission medium.

Research is being carried out into the possibility of using optical waveguides in the form of glass fibres which could carry modulated light signals. Light is a form of electromagnetic radiation, and infrared light at frequencies of the order of 100 THz (1THz $= 10^{12}$Hz) would probably be used to make vast bandwidths available. Optical waveguides differ from other bounded media in that electrical conductors are not involved.

1.4.2 Radio

Radio transmission channels use radio wave propagation between transmitter and receiver antennae. Both types of antenna can be **directional** or **non-directional.** A non-directional transmitting antenna radiates energy uniformly in all directions. It might be used with a public broadcast transmitter which is centrally located in the region it serves. A non-directional receiving antenna is equally sensitive to radiation from all directions and might be used with a general purpose radio receiver so as to make available a wide range of programmes transmitted from many different locations.

A directional transmitting antenna radiates more energy in some directions than in others and might be used for point-to-point communication. It would then be aimed so that a maximum amount of energy is directed towards the receiver.

Directional receiving antennae are more sensitive in some directions than in others. They are used, for instance, with television receivers which operate on a set of frequencies allocated to transmitters whose antennae share a common location. Directional receiving antennae have directions of minimum as well as maximum sensitivity, and it is sometimes more effective to point an antenna so that it is least sensitive to an interfering signal, than it is to point it so that it is most sensitive to the wanted signal. For instance, television reception may be affected by 'ghosts' due to waves reflected from tall buildings or hills. These waves take longer to reach the antenna than the main signal and can produce a very objectionable fainter second picture displaced from the main one. This ghost picture can sometimes be avoided by rotating the antenna so that it is least sensitive to the reflections causing the trouble.

Figure 1.5 shows a polar diagram of the type used to indicate the directional properties of an antenna. If the antenna is used for transmitting, the diagram shows the relative amount of energy radiated in each direction. If the antenna is used for receiving, the diagram shows how the sensitivity varies with direction. An antenna has the same polar diagram whether it is used for transmitting or receiving. This is true for all types of antennae.

The ratio of the radiated energy (or sensitivity) of an antenna to the radiated energy (or sensitivity) of a standard antenna (usually a dipole)

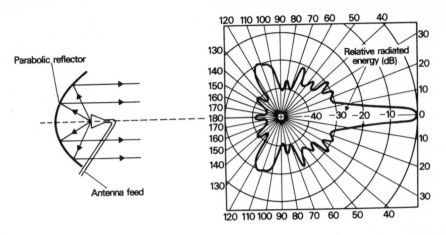

Figure 1.5 A microwave antenna and its polar diagram

connected to the same transmitter (or receiver), operating under otherwise identical conditions, is called the **antenna gain.** The radiated energy or sensitivity axis of Figure 1.5 can be calibrated in terms of antenna gain, in which case the polar diagram shows how the gain varies with direction. It can be seen from the figure that the antenna gain is greatest along the 0° reference direction and falls off sharply away from it, dropping by 25 dB over 10°. Such an antenna would be said to have a 25 dB beam width of 20°.

The gain and polar diagram of an antenna vary with frequency. The frequency giving maximum gain for an antenna depends on its dimensions. The bandwidth of an antenna is the frequency range over which the gain in the direction of maximum sensitivity decreases by a specified amount, often 3 dB, from its maximum value.

Figure 1.5 shows the radiation pattern for a microwave antenna which will be discussed later in this section. Two other antennae, a rhombic, used for long distance short wave broadcasts (4000 km and more), and a Yagi array, used, among other things, for television receivers, are shown in Figure 1.6. They have roughly similar polar diagrams in the plane of the figure. The rhombic antenna consists of a pair of horizontal wires mounted on poles. The wires form a radiating transmission line which is terminated at one end by a matching resistor. The Yagi array consists of a dipole antenna made up of two straight co-linear conductors, a set of director rods in front of the dipole, that is in the direction of maximum sensitivity, and a reflector rod behind it.

The bandwidth of a rhombic antenna depends on the length of its sides and the angles between them. An octave bandwidth, that is a 2 to 1 frequency range, can be obtained if the length of each side exceeds four times the maximum operating wavelength. The Yagi array has a smaller proportional bandwidth centred on a wavelength slightly greater than twice the total length of the dipole.

31

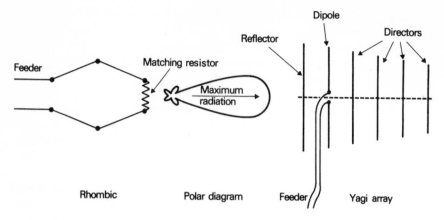

Figure 1.6 A rhombic antenna and a Yagi array. They have roughly similar polar diagrams in the plane of the figure

The orientation of a transmitting antenna affects the polarisation of the electromagnetic waves it radiates. The plane of polarisation is the plane of oscillation of the electric field in the wave. In the case of a dipole and a Yagi array, the plane of polarisation is parallel to the length of the conductors. Thus if the dipole is horizontal, the waves are horizontally polarised. The gain of a receiving antenna varies with the polarisation of the incident waves. A horizontal dipole gives maximum gain when the incident waves are horizontally polarised, and minimum gain when they are vertically polarised.

In the design of radio links, the plane of polarisation is sometimes chosen so as to minimise interference with, and from, other transmissions over the same frequency band. Thus if an existing transmission uses vertical polarisation, horizontal polarisation would be chosen for a new link.

Radio waves are unbounded in so far as they are propagated into the atmosphere. However, in the majority of cases, they are effectively bounded by the surface of the earth and by various conducting layers of ionised gases in a region called the **ionosphere,** some 50 to 400 km above the surface of the earth. Different frequency bands are reflected by different layers. It is even possible for radio waves to travel several times round the earth by successive alternate reflections from the ionosphere and the earth's surface.

At very low frequencies (v.l.f.) the waves travel round the earth as if they were in a waveguide bounded by the earth's surface and the ionosphere. Various effects contribute to this mode of propagation. They include diffraction at the earth's surface and refraction at the various layers in the ionosphere. The v.l.f. range is mainly used for navigation and military applications.

The familiar long (30 kHz-300 kHz) and medium (300 kHz-3 MHz) wave bands are known as the low frequency (l.f.) and medium frequency (m.f.) bands. They are used for radio broadcasting. They propagate by what are known as surface waves which are guided over the curvature of the earth, much as electromagnetic waves can be guided by a transmission line. Propagation in the m.f. band can also take place by sky waves, that is waves reflected from the ionosphere. Sky-wave propagation varies with ionospheric conditions, which show regular (daily and seasonal) and irregular fluctuations. Sky waves allow communication over longer distances but they are subject to fading, that is variations of signal strength due to ionospheric fluctuations. Sky-wave propagation improves at night but this is of little advantage in practice because it enables distant stations to be received at the same frequencies as local stations, and worsens the interference in the already overcrowded m.f. band.

Propagation in the short wave, or high frequency (h.f.) band of 3 MHz-30 MHz is also by sky wave. This band is used for broadcasting, mainly over long distances, for maritime or aeronautical communications and for long distance speech and teletypewriter links. Different frequencies have to be used at different times to avoid fading. The frequencies are determined from forecasts of ionospheric conditions based on observations of daily and seasonal fading patterns and also of sun spot cycles which have a marked effect on ionospheric conditions. The h.f. band is very overcrowded and subject to a great deal of interference between stations operating over the same frequencies.

Short range h.f. communications (ship to shore and land mobile) are also possible through a combination of waves travelling directly from transmitter to receiver, of surface waves and of waves reflected by the ground. This combination is called ground waves.

Both v.h.f., very high frequency waves (30 MHz-300 MHz) and u.h.f., ultra high frequency waves (300 MHz-3 GHz) propagate as space waves, that is a combination of direct and ground reflected waves, over relatively short distances ranging from a few tens of km to one or two hundred km, depending on the height of the antennae and the nature of the terrain between them. Because of the limited range, these frequency bands are relatively free of interference from distant stations. The v.h.f. band is used for high quality f.m. radio broadcasts where the relatively large bandwidth available is traded for an improved signal to noise ratio. A radio frequency (r.f.) bandwidth of about 200 kHz is used for an audio signal bandwidth of 15 kHz. Both the v.h.f. and u.h.f. bands are used for television broadcasts.

The antenna shown in Figure 1.5 is similar to those used for radar, radio telescopes and line-of-sight microwave radio link towers. It consists of a conducting parabolic dish with a relatively small antenna at its focus. This small antenna points into the dish which acts as a reflector and produces a narrow beam. A 3 dB beam width of about 1° can be obtained from a dish having a diameter of 60 wavelengths. A smaller dish would produce a wider beam.

Highly directional antennae, such as that of Figure 1.5, are useful for point-to-point radio links. The narrowness of the transmitted beam ensures that a minimum of power is radiated in directions other than that of the receiver antenna. This reduces interference with other systems operating in the same band and increases the privacy of the link. The concentration of radiated energy, that is the high antenna gain, in the required direction also allows transmitter power to be kept down.

Antennae several tens of wavelength in size become very unwieldy for wavelengths exceeding a few centimetres. They are therefore mainly used at centimetre and millimetre wavelengths, that is, microwaves. The term microwaves is often applied to radio frequencies above 1 GHz (the upper limit is not usually defined precisely and merges into the infrared region of the electromagnetic spectrum). The microwave region therefore includes some of the u.h.f. band. It also includes the super high frequency band (s.h.f.) of 3-30 GHz, and the extremely high frequency band (e.h.f.) of 30-300 GHz, sometimes called millimetre wave band.

Space wave propagation of narrow microwave beams is used for point-to-point communication over broad bands. An important application is the use of microwave links in telephone and telegraph systems. Antennae are mounted on hills or on special towers so as to give a repeater spacing of the order of 40 km. Each repeater consists of a receiver and a transmitter for each direction along the path. Each receiver antenna is located on a line of sight path from the previous transmitter. Microwave telephone links normally use f.d.m. CCIR recommendations exist for systems at 2 and 4 GHz with up to 1800 telephone channels, and at 6 GHz with up to 2700 channels. A capacity of 1800 telephone channels is adequate for one colour television channel.

One increasingly important use of narrow beam microwave transmission is with satellites which can operate as repeaters for long distance telephone, television and data links.

Tropospheric scatter propagation has been used, mainly in the u.h.f. and s.h.f. bands, to provide reliable wide-band radio links over several hundreds of kilometres. The **troposphere** is the region of the atmosphere extending from the ground to a height of 10 to 16 km. Unlike the ionosphere there is practically no ionisation of air molecules in this region. However, a beam of radio waves crossing the troposphere is scattered by mechanisms which are not yet fully understood. Figure 1.7 shows the relevant features of a

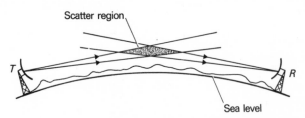

Figure 1.7 A tropospheric scatter link

tropospheric scatter link. The transmitter T and the receiver R both have narrow beam antennae. The beams intersect in the scatter region where some of the energy in the transmitter beam is scattered towards the receiver. Only a small fraction reaches the receiver and relatively large transmitter powers are needed (tens of kW compared with about 10 W for standard line-of-sight microwave repeaters). The increase in transmitter cost is justified in very inaccessible regions, such as the Arctic or on oil drilling rigs in the North Sea.

1.5 TWO SELECTED SYSTEMS: TELEPHONE AND TELEVISION

Telephone and television systems have been chosen for detailed study in the remainder of this book because they can be used to convey many important telecommunication systems ideas and also because they are very different, which is why they are treated very differently. For instance, a central theme with telephone systems is the efficient deployment and use of expensive transmission and switching plant, with many choices possible in all parts of the system: use of analogue or digital signals, various combinations of t.d.m. and f.d.m., various combinations of local and trunk exchanges, and so on.

In the case of television, the initial choice of a particular system determines the essential features of both transmitters and receivers, and most of the cost of the system is in the receivers which must not be too expensive for the viewing public. As a result, most of the parameters of television systems are chosen to give the simplest possible receiver design consistent with meeting minimum picture quality requirements, determined from perception data and user trials. On this basis, most of the treatment of television systems centres on receiver design.

Chapter 2

Switched telecommunication systems: system structure

INTRODUCTION

In this chapter we are going to look at some of the general features of switched telecommunication systems, that is systems in which a means of passing information from one terminal to another is only made available as required. This chapter will be based on a public telephone system as an example of a switched system. You should bear in mind, however, that many other switched systems have a similar structure to a telephone system. The main differences lie in the design of the terminal, and the bandwidth requirements for conveying information between the terminals. For example, a telex system for conveying alphanumeric information between two tele-typewriters has terminals which do not in the least resemble a telephone terminal, and the bandwidth required to convey the alphanumeric information is 120 Hz as compared with 3100 Hz for a telephone connection. Nevertheless, the structure of a telex system closely resembles that of a telephone system.

A switched telecommunication system is composed of three basic elements: terminals, transmission links, and exchanges. A system such as the British telephone system might contain, typically, 20 million terminals, 21 million transmission links, and 6600 exchanges. What we are going to look at in this chapter is the way in which these elements can be brought together to form an efficient and economic system which enables each user to communicate with any other user whenever he wishes.

2.1 TYPES OF SWITCHING

When you make a telephone call you normally have audible contact with the other person throughout the call, that is there is a two-way transmission channel between the two telephone terminals which is unbroken throughout the call. You can hear the room noises at the other end when the person you are in contact with is not speaking. A system which provides this kind of transmission channel is said to be **circuit switched,** because a transmission channel, or circuit, is established between the two terminals by switching together a number of transmission links, and this circuit remains established throughout the call.

In certain switched data systems it is usual to transmit blocks of data in individual packets to the required terminal. The system conveys each data packet, which is simply a long string of bits, to the required terminal, by sending it over a number of transmission links towards its destination. The links used are not in any way allocated to a particular call. As soon as one data packet has passed over a link, another packet, destined for a completely different terminal, can pass over the same link. A system which works in this way is described as a packet switched or **message switched** system. If such a system is used in a conversational mode, the two users will exchange data packets, or messages, each of which is treated by the system as a self-contained entity, and conveyed from one terminal to the other link by link. There is no continuously available transmission channel between the terminals.

Some telegraph systems work on the message switching principle, but the time taken to deliver each message is so great as to make its use in the conversational mode almost impossible. What happens in such a system is that the system is designed so as to make heavy demands on its transmission links, thus utilising them more efficiently. As a result, messages have to be placed in queues throughout the system, where they await the use of transmission links for conveyance from one point to the next. These queues normally consist of messages held in electronic stores. The total time spent queuing may amount to several hours, so that to carry on a conversation by exchanging telegrams would literally take all day. Message switched systems involving substantial queuing delays are therefore basically one-way, or non-conversational, systems. Message switched systems in general, and in particular message switched systems where the queuing delays are appreciable, are sometimes also referred to as **store and forward systems,** since messages are stored at the exchanges and forwarded once a link becomes free.

The extensive use of time sharing links, as in a packet switched data system or a telegraph system, is carried out by a special type of exchange. A distinction can thus be drawn between an exchange of this type, that is a **message switched exchange,** and an exchange which establishes a definite two-way transmission channel between the terminals, that is a **circuit switched exchange.** Circuit switched exchanges are used in telephone systems, and also in most systems involving direct person-to-person communication. Although a telephone system will be used as an example throughout this chapter, most of what will be said about the structure of the system applies to other circuit switched systems. It also applies to most message switched systems; it is only the detailed design of the exchanges which is different in message switched systems.

2.2 BASIC SYSTEM STRUCTURE

We shall now look at how a system can be structured so as to make interconnection of any two terminals possible. If there are N terminals in the system then there are $N(N-1)/2$ transmission channels that might be needed between all the possible pairs of terminals. If the system contains only a few terminals as, for example, does an intercom system in a house or office, it is possible to have $N(N-1)/2$ pairs of wires to act as transmission channels between each terminal and each other terminal. The required channel to one of the other $N-1$ terminals can be selected at each terminal by means of a set of switches. However, if the system contains more than a few terminals, such an arrangement becomes impracticable.

What is done instead is to concentrate the switching machinery at a number of central points, thus forming exchanges or, as they are called in North America, offices. Each terminal is linked to one exchange by a transmission link, referred to as a **local line,** and exchanges are interlinked by a number of transmission links called **trunks.** Each exchange has, typically, a few hundred to a few thousand local lines connected to it. In a system such as the British telephone system this leads to there being about 6 200 exchanges, each one serving an area of a few square kilometres.

To make it possible to connect each terminal to any other terminal, each exchange could be linked by a number of trunks to every other exchange. The number of trunks between any two exchanges would depend on the likely number of concurrent calls between those exchanges. The several trunks between two exchanges are referred to collectively as a **trunk route.** To fully interlink N exchanges would require $N(N-1)/2$ trunk routes. Just as to fully interlink terminals with transmission links becomes impracticable when there are more than a few terminals, so fully interlinking exchanges becomes impracticable once there are more than a few exchanges. In the British telephone system, for example, full interlinking of exchanges would require over 19 million trunk routes.

What is done in practice is to connect trunks from each exchange to a number of exchanges which can connect trunks to one another. These are called **trunk exchanges.** The exchanges to which local lines are connected are called **local exchanges.** So, a call from one local exchange to another might have a transmission channel which consists of a local line, connected by the local exchange to a trunk, connected by a trunk exchange to another trunk, and connected by the second local exchange to the required local line. In fact, some local exchanges, between which a substantial number of calls are made, have a direct trunk route between them. But in addition to this, the system is divided into a number of areas, and all the local exchanges in each area (typically 10 to 20) have a trunk route to a **primary trunk exchange** serving that area. The primary trunk exchange is said to be the **parent** of each of the local exchanges in the area. Primary trunk exchanges, between which a substantial number of calls have to be connected, normally have a direct trunk route between them. But in addition, the system is divided into a number of regions, containing several areas, and all the primary trunk exchanges in each region have a trunk route to a **secondary trunk exchange,** which is parent to those primary trunk exchanges. There are trunk routes between secondary trunk exchanges where a substantial number of calls have to be connected, but in addition there are trunk routes to **tertiary trunk exchanges** each serving a zone containing several regions. In most systems the number of tertiary trunk exchanges would be small enough for full interlinking by trunk routes to be practicable; in larger systems it may be necessary to have quaternary trunk exchanges. -

The structure of a large switched system is thus hierarchical, with fewer exchanges at higher levels than at lower ones, as illustrated in Figure 2.1.

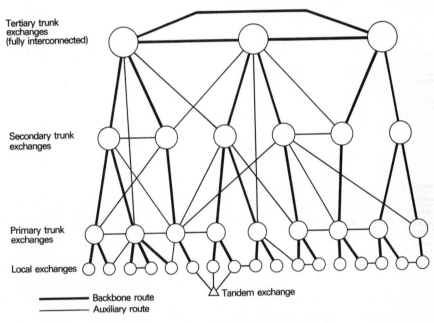

Figure 2.1 Example of an exchange hierarchy

Each exchange (except those at the highest level) has a parent exchange at one level higher in the hierarchy to which it has a trunk route. A trunk route between an exchange and its parent, or between two of the fully interlinked exchanges at the highest level of the hierarchy, is called a **final route** or **backbone route.** In addition to backbone routes there are many other routes between exchanges at the same level, and between exchanges at different levels where the higher one is not the parent of the lower one. Routes other than backbone routes are called **auxiliary routes,** and they are provided wherever there will be a substantial number of calls connected over them.

39

The part of the system consisting of the trunk exchanges and the trunks interlinking them is often referred to as the **trunk network,** the rest of the system (local exchanges, local lines, and so on) being the **local network.** As an illustration of the ratios of numbers of exchanges in the various levels of the hierarchy, the British telephone system contains about 6200 local exchanges, 370 primary trunk exchanges, 27 secondary trunk exchanges, and 9 tertiary trunk exchanges. In the U.S.A., where there are 20000 local exchanges, it is necessary to have quaternary trunk exchanges to make full interlinking at the highest level practicable. The numbers of exchanges at the levels from primary upwards are about: 1500, 200, 50, and 10.

The result of this type of hierarchical arrangement is that it is always possible, in the last resort, to connect a call between any two local exchanges by a connection which goes as follows (assuming that the tertiary level is the highest): local exchange-primary trunk exchange-secondary trunk exchange-tertiary trunk exchange-tertiary trunk exchange-secondary trunk exchange-primary trunk exchange-local exchange. The combination of trunk routes needed to connect a call is called the **routing** for the call. The routing just described, which uses nothing but backbone routes, is called a **backbone routing.** It goes up the hierarchy from each exchange to its parent, across the top of the hierarchy (except where the two local exchanges share the same tertiary trunk exchange), then down the hierarchy. Backbone routings are only used for a small proportion of calls. Most calls are connected using routings which contain one or more auxiliary routes. In fact, for most calls there will be several possible routings. The one involving the smallest number of trunk routes and keeping to the lowest possible levels in the hierarchy is called the **basic routing.** The other routings are called **alternative routings.** For a small proportion of calls a backbone routing is the one and only possible routing.

This almost completes the picture of a typical switched system structure. However, there is one more type of exchange which is sometimes used. This is a **tandem exchange,** as shown at the bottom of Figure 2.1. A primary trunk exchange is normally sited near the centre of the area which it serves. In large metropolitan areas this means that calls connected via the primary trunk exchange use trunks which pass through the densely populated parts of the area. These trunks tend to be expensive, because of the difficulties of providing underground cables in urban areas. To minimise the use of trunks into and out of the centre of a metropolitan area, a number of tandem exchanges are provided in the suburbs of the area, to connect calls between local exchanges that are outside the centre. These tandem exchanges thus take over some of the within-area functions of the primary trunk exchange.

Because the terms used for different parts of a switched system vary from country to country, it is worth noting some of the alternative names given in Table 2.1.

We shall now look at some of the operations involved in establishing a

Table 2.1 *Nomenclature comparison for the elements of a telephone system system*

Used in this book	CCITT	North American	British
Telephone terminal	Subscriber's instrument	Telephone subset	Subscriber's apparatus
Local line	Subscriber's line	Customer's loop	Local line
Exchange	Exchange	Office	Exchange
Local exchange	Local exchange	End office or central office	Local exchange
Tandem exchange	Tandem exchange	Tandem office	Tandem exchange
Trunk (between two local exchanges)	Junction circuit	Inter-office trunk	Junction
Trunk (between local exchange and trunk exchange)	Toll circuit	Trunk	Junction
Trunk (between two trunk exchanges)	Trunk circuit	Trunk	Trunk or trunk circuit
Trunk exchange	Trunk exchange	Toll office	Trunk exchange or main network switching centre
Primary trunk exchange	Primary centre	Toll center	Group switching centre
Secondary trunk exchange	Secondary centre	Primary center	District switching centre
Tertiary trunk exchange	Tertiary centre	Sectional center	Main switching centre
Quaternary trunk exchange	Quaternary centre	Regional center	—
Trunk network	Trunk network	Toll network	Trunk network or main network

transmission channel between two terminals, with particular reference to a telephone system. We shall first consider a call between two telephones connected to the same local exchange. Such own-exchange calls account for between 5 and 30 per cent of all calls. We shall then look at how the ideas that emerge from this can be applied to multi-exchange calls.

2.3 AN OWN-EXCHANGE TELEPHONE CALL

When making a telephone call, you lift the handset of the telephone and, in a very short space of time (usually a fraction of a second, though it can be several seconds) you hear dial tone in the earphone of the handset. You dial the first digit of the number you want; the dial tone stops. You dial the remaining digits and, after a wait of anything up to 15 seconds, you hear a tone. This may be number unobtainable tone, if the number you dialled was invalid or out of service; busy tone, if the telephone you are calling is in use; or ringing tone, if the called telephone is being rung. When the person you are calling answers, the ringing tone stops and you are able to converse. When you have finished, you replace the handset and the call is over. In Britain, if the handset of the called telephone is replaced during an own-exchange call, the call remains set up. So, if the handset of the called telephone is replaced and picked up again, conversation can continue. The call is not cleared down until the caller clears, that is until he replaces his handset. This arrangement is called **calling party clear.** In some cases the call is cleared down if either the caller or called party clears. This is called **first party clear.**

When a call is being set up or cleared down, the exchange is, in effect, acting upon instructions given to it by the user; the exchange is being remotely controlled by the user from his terminal. The instructions given by the user consist of a number of signals generated by the dial and by a switch inside the telephone terminal which is operated by the handset rest. The exchange indicates its response to the user's instructions by means of further signals, such as busy tone, returned to the caller. All these signals are known as **local signals** and the passing of these signals between the terminal and the exchange is called **local signalling.**

The processes involved in setting up and clearing down a call can be described in the following way. At each point during a call there can be said to exist a particular **state,** for example the called-telephone-being-rung state. Each time one of the users does something, such as dialling a digit, a local signal is sent to the exchange. This initiates a certain activity in the exchange which can have either or both of two possible results: first, a transition may occur from one state to another state and, secondly, a local signal may be sent to one or both terminals. In addition, the activity will normally involve some internal operation in the exchange, such as connecting the two local lines together. For example, consider the sequence of events illustrated in Figure 2.2. At the start, when there is no call, we can say that the **idle** state, State 0, exists. When the user picks up his handset, this generates a local signal which will be referred to as the **call-request signal.** The receipt of this signal by the exchange, as indicated by the event box on the diagram, initiates an activity in the exchange, as indicated by the activity box. In addition, a local signal is sent to the user's terminal, as indicated by the signal-sent box. This signal is, in the case of a telephone system, the start of dial tone. Because dial tone is a specific example of this particular signal, a more general term, the **proceed-to-send signal,** is used to describe it here. Once this signal has been sent, the call enters State 1, the **awaiting digits** state.

When the first digit is dialled by the user a **digit signal** is sent to the exchange. This initiates another activity in the exchange, which includes storing the digit and, in the case of a telephone system, removing the dial tone. The call enters State 2 at this point. When further digits are received, an activity is initiated which involves storing the digit and deciding whether or not a complete number has been dialled. This decision is indicated on the diagram by a decision box. Note that, in this example, no state transition occurs while the digits are being received; after each digit the call is in the same state, State 2.

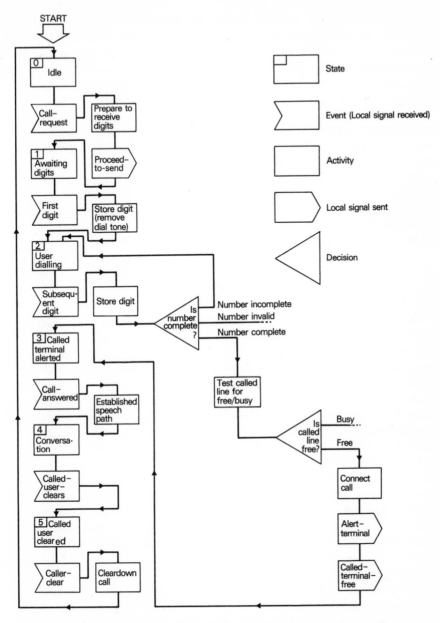

Figure 2.2 State-transition diagram for part of a telephone call

When the final digit is dialled, the exchange recognises that the number is complete and a further activity is initiated. The state of the called terminal (free or busy) is tested and, if it is free, the two local lines are connected together. Two local signals are then sent, one to the calling terminal and one

43

to the called terminal. The one sent to the calling terminal is given the general name **called-terminal-free signal.** In the case of a telephone system this is the start of ringing tone. The signal sent to the called terminal is the **alert-terminal signal.** In a telephone system this is the start of an interrupted a.c. signal (typically 75 volts, 17 Hz) which rings the bell in the telephone. The call is then in State 3, the **called terminal alerted** state.

When the called user lifts his handset, this generates a local signal, the **call-answered signal.** This initiates a further activity which involves establishing a speech path between the two terminals. In a telephone system this will also involve stopping the ringing tone and bell-ringing current. It may also involve commencement of a billing procedure to charge the caller for the call. The call is then in the **conversation** state, State 4.

For the clearing down of the call it has been assumed that calling-party-clear is used. When the called user replaces his handset a **clear signal** (called-user-clear) is sent to the exchange; this causes a transition to State 5, but no further action. However, when the caller replaces his handset, generating a clear signal (caller-clear), the cleardown activity is initiated. The outcome of this is that the idle state, State 0, is returned to, once every piece of equipment associated with the call has been returned to its idle condition.

The diagram shown in Figure 2.2 is called a **state-transition diagram.** It represents the functions performed by the exchange in a concise and useful manner. The diagram as shown in Figure 2.2 is incomplete. At each state there are possible events, other than those shown, which can occur. For example, the caller may decide to replace his handset before he has finished dialling, or the called telephone may be busy. In a more complete state-transition diagram, such as that shown in Figure 2.3, all the expected events are shown underneath each state box, each one leading to a different activity. Because the caller can clear at any point during the call, and because this will always initiate the same activity, namely the clearing down of the call, the caller-clear event box is replaced by a circle. The activity to be initiated is indicated by a corresponding circle at the top of the diagram. This makes the diagram much clearer. In fact, the circle, placed underneath the event boxes, indicates not just the action to be taken on receipt of a caller-clear signal, but the action to be taken on receipt of any signal not specified by the event boxes. Thus, cleardown is initiated if either a clear signal is received or an unexpected signal is received. This ensures that an unexpected series of events, perhaps caused by a fault, does not cause the exchange to become locked up.

In Figure 2.3 a number of possible events are illustrated which are not in accordance with the normal flow of a call. For example, the result of the called user lifting his handset again after replacing it, and thus generating a second call-answered signal, is shown under State 5. The result of this is that the call returns to State 4. The results of the called number being invalid or busy are also shown; in these cases the call may enter State 6 (number unobtainable signal sent) or State 7 (busy signal sent).

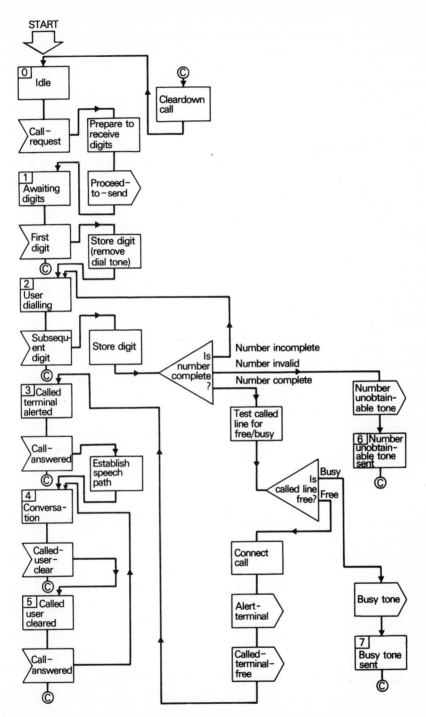

Figure 2.3 More complete state-transition diagram

Further details of state-transition diagrams and their use in telephone system design are given in Reference 34.

2.4 LOCAL SIGNALLING

The central role of local signalling in the setting up of a call should be clear from the above description. We are therefore going to look briefly at the actual form of the local signals in a telephone system.

In most telephone systems the local line consists of a pair of wires between each telephone and its local exchange. Figure 2.4 shows the parts of the telephone terminal, and the exchange, concerned with local signalling. The part of the diagram showing the local signalling circuit in the exchange is not a practical one. In practice the various parts of this circuit are located in different pieces of equipment and are connected to the local line by

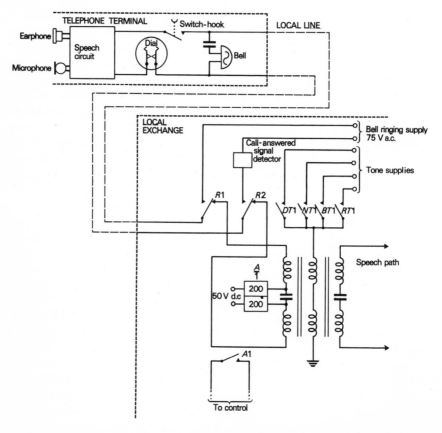

Figure 2.4 The principle of loop-disconnect signalling. The coils of relays *DT*, *NT*, *BT*, *RT*, and *R* are not shown. The call-answered signal detector has a means of informing the control of receipt of a signal

intermediate switching equipment. In case you are unfamiliar with the method of representing a relay used for relay A in this diagram, the convention is as follows. The coil of the relay is shown as a box. The figure inside the box is the resistance of the coil or coils in ohms. The letter or letters beside the box form the label used to refer to the relay, and the figure under this is the number of contact sets the relay has (so you know how many to look for in the diagram). The contacts of the relay are labelled with the letters plus a number. They may be put anywhere on the diagram; they do not have to be near the box representing the coil. (This particular convention for drawing relays is known as the detached contact convention.) The contacts are shown in the position when no current is flowing through the coil, that is when the relay is not operated. Relay A is, in fact, a double-coil relay, with two electrically separate coils, each with a resistance of 200 Ω. For the sake of clarity only the contacts of the other relays have been shown. We shall look at these relays later.

Associated with the handset of the telephone is a switch whose contacts close when the handset is picked up. This is called the **switch-hook,** a term which originates from the early days of telephones when the earpiece was separate from the microphone and was hung on a hook at the side of the instrument, rather than being placed on a cradle. (The microphone was part of the main body of the instrument.) When the handset is picked up, the switch-hook connects the speech circuit of the telephone across the local line. (The details of the speech circuit will be described in Chapter 3.) The speech circuit has a d.c. resistance of the order of 100 Ω and its connection to the line causes d.c. to flow in the local line. The start of this d.c. acts as a call-request signal on originating calls, that is on calls from the terminal. When the handset is replaced the d.c. path is broken. The cessation of the d.c. acts as the clear signal on both outgoing and incoming calls.

When the dial is pulled round and released, a series of pulses is generated by a pair of contacts which break the d.c. path through the terminal; the pulses are thus disconnection pulses. The number of pulses corresponds to the dialled digit. In Britain, one pulse is the digit 1, two pulses 2, and so on, with ten pulses for 0. In a few countries different codes are used, depending on whether the digits are in ascending or descending order round the dial, and where 0 is placed. The pulses are sent at a rate of about 10 pulses per second in most telephone systems, though 20 pulses per second are used in a few cases. The ratio of the duration of the disconnections to the time in between is nominally 2:1 in many systems, but some systems use other ratios. The dial also contains extra contacts which are closed as soon as the dial is pulled off-normal. These contacts short-out the speech circuit while the dial is operating, so as to prevent clicks in the earphone. The pulsing contacts of the dial are actuated by a cogged wheel as the dial is pulled back to its rest position by a spring, the speed of return being controlled by a governor. The end of one digit and the start of the next is indicated by an absence of any pulses for at least 200 ms. This **inter-digit pause** is guaranteed by a part of the dial's motion producing no pulses, and also by the time taken for the user to pull the dial round and release it again.

The basic features of the circuits involved in local signalling at the exchange are illustrated as follows. (Remember that Figure 2.4 is not a practical circuit.) Voltage is fed to the local line through a double-coil relay, labelled A, to facilitate signalling and provide power for the microphone in the speech circuit of the telephone. The transformer provides a path for a.c. signals, that is, for speech signals, but isolates the local line from the rest of the exchange as far as d.c. is concerned. The capacitor across the relay provides a low impedance path for speech signals so that the full speech voltage from the line is fed to the transformer. Whenever there is d.c. flowing through the local line, relay A operates. The contacts of relay A, labelled $A1$, feed the signals extracted from the local line to the part of the exchange which exercises control over switching operations. For the moment this part of the exchange will simply be referred to as the **control.** The call-request signal is presented to the control as the closing of contacts $A1$. The dial pulses are presented as momentary openings of contacts $A1$. The clear signal is the opening of contacts $A1$ for a period considerably longer than the duration of a dial pulse, say 200 ms.

There are five signals that the control may need to send to the terminal: dial tone, number unobtainable tone, busy tone, ringing tone and 75 volts a.c. to ring the bell on an incoming call. To send any of these signals the control applies a voltage to one of five relay coils DT, NT, BT, RT and R. For the sake of clarity these have not been shown on the diagram; only the contacts of these relays are shown. Contacts $DT1$, $NT1$, $BT1$, and $RT1$, are arranged so as to connect tone supplies to a tone-injection winding of the transformer. Thus, when one of these relays is operated by the control, the appropriate tone is sent to the terminal. When relay R is operated by the control, the local line is connected by contacts $R1$ and $R2$ to a 75 volts a.c. supply which rings the bell in the telephone. The purpose of the capacitor in the terminal is to provide a path for the a.c. ringing current but prevent d.c. flowing through the bell and interfering with the d.c. signals generated by the dial and switch-hook. The call-answered signal, which consists of the connection of the speech circuit of the terminal to the local line when the handset is picked up, is detected by a call-answered signal detector, which is in series with the 75 volts ringing current supply. This detector can distinguish between the impedance of the bell plus capacitor and the impedance of the speech circuit by means of a d.c. voltage superimposed on the ringing current, causing d.c. to flow once the call is answered. The design of this detector will not be discussed here. On detecting the call-answered signal it sends an electrical condition to the control, which releases relay R so that the local line is then connected as on an outgoing call for the remainder of the call.

The method of digit signalling using disconnection pulses, as described above, is known as **loop-disconnect signalling.** The digit signals, although traditionally generated by means of a dial, may also be generated by a circuit controlled by push buttons, which generates the pulses electronically. These are generally considered to be easier to use but, surprisingly, they do not appear to reduce the number of dialling errors. The use of an electronic circuit to generate the signals does not reduce the time taken to set up a call. Because the time taken to send a digit is up to 1.2 seconds (ten pulses for 0 followed by the minimum inter-digit pause of 200 ms), the user has to wait for several seconds after keying a number while the electronic circuit pulses it out. It is not normally possible to increase the speed at which pulses are sent because this would mean redesigning the equipment in the local exchange.

Another method of digit signalling, known as multifrequency tone signalling, makes it possible to send each digit as the push buttons are pressed. The push buttons, in this case, are connected to a set of oscillators which apply a pair of tones to the local line each time a button is pressed. These tones are detected at the exchange, and the digit thus determined. The tone detection circuits at the exchange can recognise the tones in, typically, 33 ms. The tones agreed by the CCITT for use in this type of local signalling are shown in Figure 2.5. Each row and column of the set of push buttons is allocated a tone. The two tones sent when a button is pressed are those for the row and column in which the button lies. The d.c. power supply for the oscillators in the telephone is drawn from the local line and the sending level

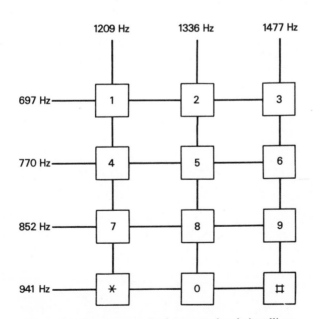

Figure 2.5 Tones for multi-frequency local signalling

for the tones is comparable with that of normal speech signals from the microphone. The d.c. signals generated by the switch-hook are the same as with loop-disconnect digit signalling. Note that there are two extra digits available when multifrequency signalling is used. The CCITT have recommended that they be labelled ✳ and # . These will be used for certain special services, such as controlling the automatic transfer of calls to other telephones. Only certain types of exchange are able to respond to the increased speed of signalling that multifrequency tone signalling brings about.

2.5 NUMBERING SCHEMES

Before going on to look at how calls involving several exchanges are set up, it is convenient at this point to examine the way that the user makes his wishes known to his local exchange by means of the digit signals. The digits which the user sends to the exchange using a dial or push buttons define either the telephone to which he wants to make a call or some special service which he wants, for example, an operator or a recorded information service, such as the speaking clock. Each telephone and each service must be ascribed a set of digits, or **address** as it is called, and the scheme by which addresses are allocated is referred to as a **numbering scheme.**

The first requirement that a numbering scheme must meet is that the initial digits of one address should not coincide with another complete address. For example, if 294890 is an address, 2948 cannot be used as another address and neither can 2, 29, 294 or 29489. This is simply because the digits of an address are sent one after another, rather than all at once. If 2948 were an address, then a user dialling 294890 would be connected to 2948 as soon as he had dialled those digits, unless a special code or digit (for example #) is used to indicate the end of the address. This is because without an end of dialling signal, the exchange has no way of knowing that more digits are to follow.

One of the simplest ways of devising a numbering scheme is to give each telephone on an exchange a different number of the minimum possible length, taking into account the number of telephones. For example, on an exchange with up to 10000 telephones, 4-digit numbers would be needed. Users making a call from one exchange to another then have to dial a code in front of the required number to indicate which exchange it is on. These codes can be made to define the routings to be used. The digits dialled thus give the local exchange its instructions in a very direct form. However, this leads to a situation in which the digits dialled on a call to a particular exchange vary according to where the user is calling from. Also, for calls where the routing involves a number of intermediate exchanges, the code used to define the routing has to be a long one. This is an example of forcing the user to fit the system, rather than fitting the system to the user. This arrangement of codes based on the routing is still used for local calls in some parts of Britain, although it is now being phased out.

A preferable arrangement, which takes account of the needs of the user, is to allocate longer, but unique numbers to each telephone over a given geographical area. This can be done by allocating certain groups of initial digits to each exchange. For example, if the digits 294 are allocated to a particular exchange, then all numbers on that exchange are of the form 294XXX, where XXX are any three digits. Numbers of the form 295XXX would be on a different exchange, and so on. Thus, when a user dials 295123, for example, his local exchange has to examine the initial digits 295 and decide to which exchange the call is being made and what routing is required to reach that exchange. The digits 295 are said to be **translated** into the required routing. The user therefore dials an address which indicates the destination, instead of the routing.

Within the area in which such a numbering scheme is operative, the numbering of telephones on different exchanges is thus linked, and such a scheme is called a **linked-numbering scheme.** The result is that, within the area, users have to dial only the number of the required telephone; no special dialling codes are needed in front of the number. This principle can be extended a stage further, by prefixing the numbers in each area of a country by a group of digits which identifies that area. The country is thus split into what are called **numbering plan areas,** each of which is identified by an **area code.** The complete number, formed by prefixing the basic number by an area code, is called a **national number.** For example, there might be a number 294890 in many areas, but there would only be one national number 712-294890, namely the one in area code 712. The national number 713-294890 would be in area code 713.

It is normal practice to instruct users to dial just the basic number, for example 294890, to call that number within the numbering plan area. This basic number is called the **local number.** For inter-area calls the complete national number is dialled. This arrangement is sometimes referred to as **two-tier dialling.** Because it is necessary for an exchange to distinguish between the national number 712-294890 and local number 712294, a national number prefix is normally used to warn the exchange that the number to follow is a national number. Typical prefixes are 0 (used in Britain and many other countries) and 1 (used in parts of North America). In other parts of North America the need for a prefix is eliminated by making sure that the combinations of digits used as area codes are not the same as the initial digits of any local number; the area codes consist of three digits and can be distinguished from the initial digits of a local number by the second digit of an area code always being a 0 or a 1.

The principle of two-tier dialling and area codes is extended one stage further for the purposes of international calls. Each country is allocated a **country code** and this is combined with the national numbers of each country to form **international numbers,** just as area codes are combined with local numbers to form national numbers. So, for calls between countries, the complete international number is used. For example, national number 712-294890 in country 44 (Britain) becomes international number 44-712-294890. Once again, it becomes necessary to use a prefix to warn the exchange that the number to follow is an international number. Typical international prefixes are 00 (in many European countries) and 010 (in Britain). The CCITT has allocated country codes to all countries. These are one-, two-, or three-digit codes, for example, North America 1, USSR 7, South Africa 27, Britain 44, Australia 61, Luxembourg 352, Saudi Arabia 966. The CCITT has also recommended that the international prefix should never be included when quoting an international number on letterheads, since the prefix varies from country to country. Instead, a + should be placed in front of the number, for example, +44-712-294890.

Codes for services such as the operator, speaking clock, and so on, are normally standardised throughout a country. This saves the user having to look up the code wherever he goes. For convenience the codes are kept as short as possible, usually one, two, or three digits. For example: 100 (operator in Britain), 999 (emergency operator in Britain), 16 (speaking clock in Italy) and 0 (operator in North America).

2.6 MULTI-EXCHANGE CONNECTIONS

We shall now look at what happens when a call involving more than one exchange is set up. The user picks up his telephone handset, (or performs whatever operation is necessary in the system in question, to generate a call-request signal). The local exchange returns a proceed-to-send signal, such as the connection of dial tone. The reason for this is that the exchange is not immediately ready to receive digit signals. Equipment to receive digits must be connected to the local line. The proceed-to-send signal indicates that this has been done. The user then sends the appropriate digit signals that form the address of a terminal, which in this example is connected to a different local exchange. The action taken by the exchange is as follows.

First, the exchange translates the first few digits of the address into a routing, or at least a part of one. It then examines the trunk route to the next exchange on the routing, to find a free trunk in that route, and connects the local line of the calling telephone to this trunk. Next, the local exchange must request the exchange at the other end of that trunk to perform a switching operation to connect this trunk to a free trunk in the next route in the routing. It does this by means of a process of **inter-exchange signalling.** Just as the user can remotely control his local exchange, by means of local signalling, so one exchange can remotely control another, by means of inter-exchange signalling.

To illustrate how this happens, we shall consider a telephone call between two terminals on different local exchanges, between which there is a direct trunk route. When the first exchange has selected a trunk to the second one it must do two things. First, it must inform the second exchange that a call is about to be connected over that trunk and, secondly, it must indicate what switching operations are required. This can be done by a simple extension of the local signalling process. The first exchange sends a call-request signal to the second one, to indicate that a call is required. When the second exchange is ready to deal with this, it returns a proceed-to-send signal. (In some cases this is not needed; the second exchange may be designed to be ready to receive digits within a certain time from receipt of the call-request signal.) The first exchange repeats as many digits as necessary of the address supplied by the user. The second exchange connects the trunk from the first exchange to the required local line, just as though the trunk were a local line from one of the telephones on that exchange. The called-terminal-free signal (the start of ringing tone) is sent back from the second exchange to the first one, and hence to the caller, and the alert-terminal signal (the application of bell ringing current) is sent to the called terminal by its local exchange.

Thus far the signalling processes are very similar to those on an own-exchange call. When the called user answers, a call-answered signal passes from the called terminal to its local exchange. This must somehow be passed back to the first exchange so that a billing procedure can be initiated there. As far as the user making the call is concerned, the cessation of ringing tone acts as the call-answered signal, informing him that the called user has picked up his handset. However, this signal is not easily recognised by the first exchange, so it is backed up by another call-answered signal, sent from the second exchange to the first as part of the inter-exchange signalling process.

At the end of the call, when a clear signal is sent from the calling terminal to the first exchange, a clear signal is forwarded to the second exchange so that the connections at both exchanges are cleared down at once.

When more than two exchanges are involved in the setting up of the call a similar procedure applies. The first exchange sends signals to the second exchange which, acting on the routing information contained in the digit signals, identifies the next exchange in the routing and sends similar signals to that one. This process is repeated until the call is extended to the destination local exchange. The main difference between the particular case of a call connected directly between two local exchanges, and a call involving more than two exchanges, lies in the information contained in the digit signals. This information must define not only the terminal to which the call must be connected, but also the routing which is to be used to get to the destination local exchange. There are two ways in which one exchange can get another to select the next trunk route.

The first way of routing a call is for one exchange, either the originating local exchange or its primary trunk exchange, to determine almost the whole of the routing by translating the address supplied by the user. For example, if the call is between two numbering plan areas, the originating local exchange may forward a complete national number, as dialled by the caller, to its primary trunk exchange. The primary trunk exchange can then examine the area code in the national number and determine the required routing to get to the primary trunk exchange of the numbering plan area which has that area code. This routing may then be specified by, say, six digits. These six digits consist of two groups of three digits, each of which represents an instruction to another trunk exchange. These digits are sent before the local number, in place of the area code. The first trunk exchange examines the first group of digits. This tells it which trunk route is next in the routing. The call is switched through to a trunk on that route, and the digits, minus the group which it has acted on, are forwarded to the next trunk exchange. This next trunk exchange acts on the second group of routing digits in a similar manner, and thus extends the call to the required primary trunk exchange. The second group of routing digits is omitted this time, so that only the local number is sent to this primary trunk exchange. This primary trunk exchange, which is the one allocated to the numbering plan area to which the local number relates, can then translate the initial digits of the local number to identify the destination local exchange and thus extend the call to it. The local number is forwarded to the destination local exchange and the final stage of the setting up of the call proceeds just as though it had come direct from another local exchange, as in our earlier example.

The point to note about this first method of routing is that a substantial part of the routing, namely that from the first primary trunk exchange to the last, is decided by a single translation process in the first primary trunk exchange. Subsequent exchanges are simply acting as slaves to this decision by treating the routing digits as explicit instructions for the selection of the next trunk route. This method of routing is referred to as **forward code routing.**

The second method of routing a call is for each exchange to send forward the complete number (local, national or international, as appropriate) to the next exchange, without changing or adding to the digits. Each exchange then performs an individual translation process to determine the next exchange to which the call should be extended. The routing of the call thus comes about, stage by stage, rather than being determined at a particular point in the system. Each exchange concerns itself only with the destination to be reached, and extends the call in the right direction. This method of routing is called **destination code routing.**

Destination code routing has two significant advantages over forward code routing. First, it requires each exchange to hold less translation information. With destination code routing, each area code (or whatever part of the address has to be translated) corresponds to a choice of one trunk route, that is the one which the call is extended over to the next exchange. But with forward code routing, each area code corresponds to a description of a complete routing. Hence, the information stored in the translating equipment is several times greater than with destination code routing. Not only does the extra storage capacity cost more, but also the difficulties of updating the information, when changes are made to the system, are greatly increased. The second advantage of destination code routing arises from the ability of each exchange to select the next trunk route from several possibilities. It was mentioned earlier that there are often several possible routings for a call. This being so, it is possible to use a routing, other than the basic one, whenever the first-choice trunk route has no free trunks on it. This technique is referred to as **alternative routing.** All that it involves is that each exchange translates the destination code into at least two possible routes for the next stage of the routing: a first-choice route, second-choice route, and so on. Each route is then examined in turn. If there are no free trunks on the first-choice route, then the second is tried, and so on. The ability to do this can reduce the total number of trunks needed on each route under certain circumstances, as will be described in Chapter 4.

2.7 INTER-EXCHANGE SIGNALLING

We are now going to look at some specific examples of inter-exchange signalling. There are a number of methods in use throughout the world. Each country has several of its own for national use, and the CCITT has defined several signalling systems for use on international telephone calls. The method chosen for a particular trunk route depends on a number of factors, the principal ones being the type of transmission links used on the route and the types of exchange at either end.

Before looking at an example of inter-exchange signalling, it may be helpful to summarise the signals which such a system has to convey. This is done in Figure 2.6, which is a **signal diagram,** showing the sequence of signals passing between the calling terminal, its local exchange, a second local exchange, and the called terminal. In principle there could be any number of trunk exchanges interposed between the two local exchanges; the signals shown are simply repeated by intermediate exchanges. Note that the called-terminal-free signal (the start of ringing tone) and the call-answered signal (the cessation of ringing tone) pass straight through exchanges to the calling terminal. However, this audible call-answered signal is backed up by a second signal which can be easily recognised by exchanges and which is repeated from exchange to exchange. In some systems the called-terminal-free signal is also backed up in this way. The clear signal, repeated forward from the originating exchange, is referred to as the **clear-forward signal.** The clear signal from the called terminal is repeated back to the originating exchange. This is referred to as the **clear-back signal.**

For simplicity, only four dialled digits have been shown in this example. Note that a call-request signal cannot be sent to the next exchange from the originating exchange until sufficient digits have been dialled for the destination to be determined. This has been taken as two digits in this example. The sending of digits forward over the inter-exchange signalling system starts off a few digits behind the sending of digits from the calling terminal. This is why there is often a noticeable **post-dialling delay** before a multi-exchange call is connected; the inter-exchange signalling has to catch up with the user's dialling. This does not apply where the need for translation is avoided by making the digits define the routing. Where this is the case the post-dialling delay may be negligible.

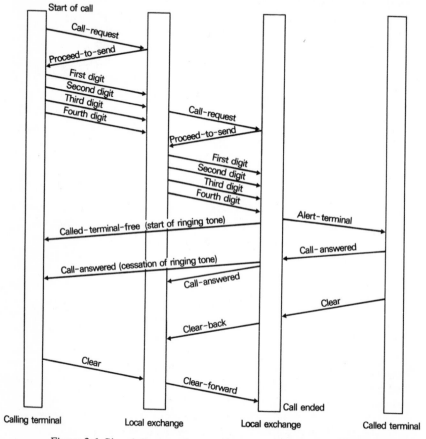

Figure 2.6 Signal diagram for a call between two local exchanges

There are a number of different methods of inter-exchange signalling. On a particular trunk route one method, or a combination of two, may be used. The principal methods are as follows:

(a) Direct current signalling. This, like local signalling in a telephone system, relies on there being a d.c. path through the transmission link itself for the conveyance of signals.

(b) In-band a.c. signalling. This does not need a d.c. path through the transmission link. Sinusoidal tones are used to convey signals over the transmission link.

(c) Out-band a.c. signalling. This uses tones outside the speech band to convey the signals. This is used mainly on older types of f.d.m. transmission links; the tones or tone, for example 3825 Hz, fit(s) in the gap between one 300-3400 Hz channel and the next. (There is normally a 900 Hz gap to minimise interference between channels.)

(d) Separate channel signalling. Channels which are separate from the transmission links are used to convey the inter-exchange signals corresponding to calls on one or several transmission links.

A widely used method of d.c. signalling is simply an extension of loop-disconnect local signalling. A simplified version of the sort of arrangement used is illustrated in Figure 2.7. In this example the trunk is used for calls in one direction only, that is calls set up from the first exchange to the second. The practice of using trunks in one direction only, and thus having separate trunk routes for the two directions, is quite a common one. It

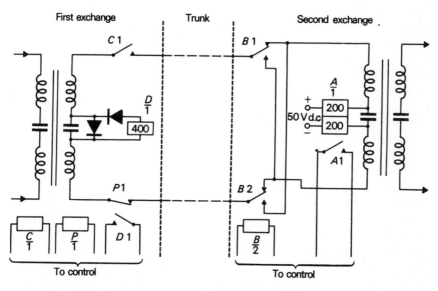

Figure 2.7 The principle of loop-disconnect inter-exchange signalling

greatly simplifies the signalling procedure and there is no possibility of calls from the two ends clashing. (Some trunk routes, however, use trunks operated in both directions.) In the first exchange, the control causes relays *C* and *P* to operate and release as required. To send a call-request signal, the control operates C. Contacts $C1$ establish a d.c. path through the transformer windings and this acts as a call-request signal. When the control operates relay P, contacts $P1$ break this d.c. path. Pulsed operation of P thus sends loop-disconnect digit signals. A clear-forward signal is simply the breaking of the d.c. path for longer than 200 ms. This is brought about by the release of relay C. (Remember that the dial pulses are much shorter disconnections, typically 67 ms.)

At the second exchange the detection of these signals, using relay A, is carried out just as with local signalling. To send signals in the reverse direction relay B is used. When B is operated by the control in the second exchange, the polarity of the power feed to the trunk is reversed. This causes relay D, at the first exchange, to operate. The operation of D closes contacts $D1$ and this acts as a call-answered signal to the control in the first exchange. When B is released, D releases. This acts as a clear-back signal. With this simple arrangement, no proceed-to-send signal is used. Instead, the first exchange sends a call-request signal, waits a certain length of time, and then sends the digit signals; it assumes that the second exchange will then be ready for them.

In North America a method of inter-exchange signalling similar to this is used in which a reversal of the power feed polarity acts as a proceed-to-send signal, and return to normal polarity acts as a call-answered signal.

There are two distinct kinds of in-band a.c. signalling in use. One is simply an a.c. version of loop-disconnect signalling in which the d.c. signals are replaced by tones. For example, using a single tone, the inter-exchange signals may be as follows:

Call-request—60 ms of tone
Digits—tone for the duration of the loop-disconnect in the d.c. system
Clear-forward—1 s of tone
Call-answered—250 ms of tone
Clear-back—250 ms of tone

Here again, there is no explicit proceed-to-send signal. Note that the 250 ms signals in the backwards direction are not ambiguous; the first exchange knows which is which, because they can only occur in alternation. The tone used is normally one which is unlikely to occur at a high power level in normal speech; accidental imitation of signals by speech during a conversation is therefore unlikely. Tones in the frequency range 2000 Hz to 2800 Hz are most suitable for this purpose.

The second kind of in-band a.c. system uses multi-frequency digit signals, rather like those used for local signalling, except that the frequencies of the tones are different, and only five or six are used instead of seven; every possible pair of tones from these five or six acts as a digit signal. The use of six tones provides five extra digit signals for special signalling facilities. Each digit may take typically 100 ms to send, as compared with an average of 750 ms (300 ms for digit 1 and 1 200 ms for digit 0, including the inter-digit pause of 200 ms) for systems based on loop-disconnect digit signals. Multi-frequency systems use similar signals to the other a.c. systems for call-request, clear-forward, and so on. It should be noted that, in principle, a.c. signalling systems using a single tone could be used to signal at a higher speed by increasing the pulse rate, though in practice this would involve substantial modification of the exchanges between which the signalling takes place.

Out-band a.c. signalling systems work on a similar principle to the in-band ones, except that the tone used is outside the speech band. Because they are not compatible with digital transmission systems they are obsolescent.

Separate channel signalling is a more recent method of inter-exchange signalling. There are two forms which it may take. In the first form, each speech channel has its own separate signalling channel. This is the case in some digital transmission systems. The second form is **common channel signalling,** in which one digital signalling channel is shared between a number of trunks. Every inter-exchange signal for a number of trunks (possibly several hundred) is represented by a distinct pattern of, say, nine bits. These are prefixed by, say, eleven bits to identify the trunk to which the signal belongs, and the twenty bits are sent over a digital channel between one exchange and the other. There are two channels, one in one direction and one in the other. Thus, all the signals for the trunks are multiplexed on-to a single two-way channel. This type of system is particularly suited to digital transmission systems, where a spare speech channel can serve as the signalling channel. Common channel signalling is also especially suited to exchanges using computer control. Between two such exchanges common channel signalling can be achieved by simply providing a digital data link between the two computers.

There are a number of practical problems with common channel signalling. To start with, it is important that the signals be protected against errors, which are much more likely to arise on a high bit-rate common channel than using conventional signalling over individual speech channels. Error protection is normally achieved by adding redundant bits to the signal messages and using these to perform parity checks on the message. A second problem is that of signalling equipment failures. If the common channel fails, a considerable number of trunks are disabled. It is therefore necessary to have a standby channel for use in the event of the main one failing.

The CCITT have defined a common channel signalling system, known as CCITT No. 6, for use on international calls. This uses 28-bit messages: 11 bits to identify the trunk, 9 for the signal itself, and 8 for error protection. The messages are sent over a 2.4 kbits s^{-1} data channel between the international exchanges in different countries. Similar systems are being developed by a number of countries for national use.

2.8 INTERNATIONAL CALLS

A brief description of the international telephone network will now be given. It is worth emphasising at this point that the most powerful influence over the development of the international telephone network has been the CCITT. The recommendations of the CCITT have led to the standards necessary to achieve satisfactory international working. These standards have had repercussions throughout the telephone systems of most countries.

The international network consists of a number of international exchanges and a network of international trunks. Each international exchange normally performs two separately identifiable functions. First, it is an **international gateway,** providing a means of connecting calls from the national system of the country to the international network. Secondly, it is a trunk exchange for calls connected between other countries. That is, a call from country A to country C may be connected from the international exchange of country A, over a trunk to the international exchange of country B, and then over another trunk to the international exchange to country C. The international exchange of country B is thus acting as an international trunk exchange, interconnecting the trunks from countries A and C. In the context of international calls it is usual to refer to exchanges which do this as **transit centres,** or CTs (CT is short for Centre du Transit).

For the purpose of routing international calls, a hierarchy of transit centres has been established, just as national networks have a hierarchy of trunk exchanges. The CCITT has defined three levels of transit centre. The highest level transit centres are fully interlinked by international trunks. These are referred to as CT1s. The lowest level centres are CT3s, analogous to primary trunk exchanges, and the intermediate centres are CT2s. The backbone routings for international calls may contain five international trunks and pass through two CT3s, two CT2s, and two CT1s.

Because of the marked fluctuations in the number of calls between different countries, owing to time differences, the use of alternative routing in the international network is extremely important in ensuring efficient use of the network. Because of this, and because of the need to make alterations to parts of the network without changing translation information throughout the world, destination code routing is used throughout the international network. One interesting feature of many of the international signalling systems is that, rather than sending the complete international number over the last trunk to the destination country, the penultimate international exchange omits the country code and uses a special call-request signal instead. This informs the exchange at the start of signalling that the call is terminating in its country. Other signalling systems use a special digit signal, at the start of the digit signals, to indicate whether the call is a transit or terminal call.

The automatic international switching network is still some way from being complete. Operator controlled switching is still used for a proportion of international calls, although the number of countries with international dialling facilities is growing rapidly. The CT1s that have been nominated by the CCITT include the international exchanges in London, New York, Moscow, Sydney, Tokyo and Singapore.

2.9 TRAFFIC

Before concluding this chapter it will be useful to introduce the topic of **telecommunication traffic.** This will be dealt with in more detail in Chapter 4 when we look at aspects of the performance of switched systems. Telephone traffic will be used as the basis for the following description, but the same principles apply equally to data traffic, telex traffic, visionphone traffic, and so on.

The planner of a road system needs to know the likely number of vehicles that will use the roads, in order to make the roads of adequate size. In a similar, but not exactly analogous way, the planner of a switched telecommunication system must match the capacity of the system to the telecommunication traffic. If the system has insufficient capacity then it will become overloaded or **congested,** and if it has too much capacity then resources will have been needlessly expended.

Whilst the behaviour of users as individuals is unpredictable, it is generally true to say that the behaviour of people, considered in large enough groups, appears regular and consistent. All forms of public service depend on this. The random behaviour of individuals is averaged out when large numbers are viewed as a whole. If we were to look at several hundred terminals connected to a particular local exchange for, say, two minutes and record the number of users making a call, the result might be as shown in Figure 2.8. Each upward step on this graph corresponds to an individual call starting and each downward step to a call finishing. You can see that, although calls start and finish more or less randomly, there is a tendency for the number of calls in progress to be fairly steady. The average number of calls, 43·5, is indicated by the dotted line.

The traffic generated by the users, and therefore carried by the equipment, can be described in terms of the number of calls. The number of calls in progress at any time may be regarded as the **instantaneous traffic.** The average value of this over a given period of time is the **average traffic,** or more commonly, just **traffic.** To indicate that this quantity describes the traffic, a dimensionless unit called the **erlang** is used. (This is named after the Danish mathematician, Agner Erlang, who developed much of the original traffic theory.) Thus, an average of 43·5 concurrent calls is described as 43·5 erlangs of traffic.

Another way of approaching the definition of traffic is as follows. If a total of n calls are made during a period of observation of T seconds, and the durations of the calls are $h_1, h_2, h_3, h_4, \ldots h_n$ seconds, the total use made of

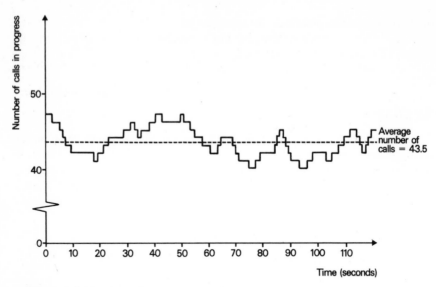

Figure 2.8 Number of concurrent calls in an exchange over a two-minute period

the system by the users in the period T can be described as $\sum_{i=1}^{i=n} h_i$ call-seconds. This is sometimes called the **amount of traffic** or **traffic volume** for the period T. The average traffic, that is amount of traffic per unit time, can be defined as:

$$E = \frac{\sum_{i=1}^{i=n} h_i}{T} \text{ erlangs} \tag{2.1}$$

Equation (2.1) can be used to derive two equivalent expressions from which the traffic might be determined using practical measurements. The average time for which calls occupy the equipment, which is referred to as the **mean holding time,** h, is:

$$h = \frac{\sum_{i=1}^{i=n} h_i}{n} \text{ seconds} \tag{2.2}$$

From Equation (2.1) it follows that the average traffic, E, for the period T, is equivalent to (nh/T). Furthermore, the average rate at which new calls appear, which we shall refer to as a calls per second, is given by (n/T) for the period T. E can therefore also be expressed as (ah). There are thus four ways of deriving a value for E:

(a) the average number of concurrent calls,

(b) the traffic volume for a period T, divided by T: $E = \dfrac{\sum\limits_{i=1}^{i=n} h_i}{T}$ erlangs,

(c) the number of calls in a period T, multiplied by the mean holding time, divided by T: $E = \dfrac{nh}{T}$ erlangs,

(d) the call arrival rate, a, multiplied by the mean holding time: $E = ah$ erlangs.

The expression

$$E = ah \text{ erlangs} \tag{2.3}$$

is a particularly useful one, which we shall be using in Chapter 4 when we look at traffic theory in more detail.

The traffic carried by a particular part of a telecommunication system, for example an exchange or a trunk route, varies throughout the day. The pattern of traffic tends to be similar during each weekday, and reaches a peak during the morning in most exchanges and on many trunk routes. In order to have a single quantity which characterises the traffic pattern for planning purposes, the average traffic during the busiest hour of a weekday is used. This is called the **busy hour traffic.** When a traffic level is quoted without any mention of the period to which it applies, you can normally assume this to be the busy hour traffic for the part of the system in question. The choice of an hour as the averaging period is to some extent arbitrary. The use of a shorter period would make it difficult to identify a busy period which is consistent from day to day and a longer period would average out the peaks of traffic too much. When a call is **offered** to a particular part of the system, for example a trunk route, there is a finite probability that there will be no equipment available for it. For example, all the trunks on a trunk route may be in use. The call is then said to be **blocked.** The user is informed of this situation by being sent a congestion signal, such as a special tone. The blocking of calls leads to the notion of the **probability of call blocking.** If, during the busy hour, C_O calls are offered and C_B of these are blocked, then the probability of call blocking, B, is given by:

$$B = \frac{C_B}{C_O} \tag{2.4}$$

It should be noted that the average traffic is always less than the number of items of equipment available to carry calls. For example, a trunk route of 63 trunks might carry 50 erlangs of traffic, so that the average traffic per trunk would be $50/63 \simeq 0 \cdot 8$ erlang. However, the number of calls at any moment may vary between 0 and 63. The probability of there being 63 calls at any given time is about 0·01. In most situations the probability of call blocking is equivalent to this, because blocking occurs when all the items of equipment are in use.

As you will see in Chapter 4, it is possible to calculate the probability of call blocking for each part of a system, given the number of items of equipment which can carry calls and the traffic which is to be offered to that equipment. To do this it is necessary to make assumptions about the randomness of user behaviour and the action taken by users when a congestion signal is received. The above figure of 0·01 for the probability of call blocking when 50 erlangs of traffic are offered to 63 items of equipment is the result of such a calculation.

2.10 MESSAGE SWITCHED SYSTEMS

Throughout this chapter it has been assumed that the structure of message switched systems is basically the same as that of circuit switched systems. The different way in which a message switched system works arises from the operation of its exchanges. However, there are a few other points of difference which must be mentioned to avoid confusion.

First, whereas circuit switched systems have trunk routes which consist of a number of distinct transmission links, a message switched system has no need for separate links. All that is required is a single means of conveying messages from one exchange to another. Messages destined for different terminals follow one another down the transmission links, so in principle there is no point in having separate links on one route, though in practice several links may be used on one route to protect against the effects of a link failure and to provide incremental growth in traffic capacity. As far as the capacity of routes in the two types of system is concerned, in a circuit switched system the transmission links have a fixed information capacity (or bandwidth) and the number of links is varied according to the traffic, but in a message switched system the information capacity of the routes as a whole is varied according to the traffic.

The second difference is that, in a message switched system, the signalling processes are not as separate from the communication processes as they are in a circuit switched system. The address of the terminal to which each message or packet is to be sent is in exactly the same form as the message itself, this normally being binary digital form. The address simply forms a prefix to the message. In order to route the message to the required destination, each exchange examines this address and selects the next route over which the message is to be sent.

The third point of difference between the two types of system is concerned with blocking. In most circuit switched systems calls are blocked if there is no equipment to handle them; to connect all calls would require very large amounts of equipment. But in a message switched system it is a basic aim of the system to deliver all messages to the required destination, no matter what happens. The user must be certain that his message will reach his destination when he sends it; he has no way of knowing whether or not the message has arrived at its destination so he cannot be expected to make repeated attempts to send the message. What happens in a message switched system, therefore, is that messages are placed in queues wherever a transmission link is not immediately available, and they are transmitted at some later time, no matter how long they have to wait. The performance of a message switched system is therefore measured in terms of the probability of messages taking more than a certain time to reach their destination, rather than in terms of probability of call blocking.

Chapter 3

Switched telecommunication systems: terminals, transmission links and exchanges

INTRODUCTION

In Chapter 2, the main elements of switched telecommunication systems—terminals, transmission links and exchanges—were treated largely as black boxes. In this chapter we are going to look inside those black boxes. As in Chapter 2, a telephone system will be used as an example. However, the basic features of transmission links and exchanges that will be described in this chapter are typical of many switched systems. To start with there is a brief description of some of the characteristics of the overall transmission channel (between one terminal and another) and ways of expressing these characteristics mathematically. Terminals, transmission links and exchanges are then examined in that order.

3.1 CHARACTERISTICS OF THE OVERALL TRANSMISSION CHANNEL

The transmission channel between two terminals which send and receive analogue signals can be described, first, in terms of its bandwidth, and secondly, in terms of the impairments of the message signals which it causes. These impairments may be due to:

(a) Attenuation.

(b) Variation of attenuation with frequency.

(c) Harmonic distortion: non-linearity of the channel.

(d) Non-linear phase characteristic: change of propagation time with frequency.

(e) Noise: thermal noise, impulsive noise (clicks and crackles), hum (mains frequency interference), crosstalk (message signals leaking through from other channels), and quantisation noise (due to the use of p.c.m. on part or all of the channel).

(f) Echoes: delayed versions of message signals produced by reflections in the system.

(g) Clipping: the cutting off of the start of each group of signals due to, for instance, the action of echo-suppressors.

For a binary digital transmission channel, the measure of the information capacity is the bit-rate and impairments of the message signals may be due to:

(i) Noise, causing single or multiple bit errors, so that a 1 is received as a 0 or vice versa.

(ii) Faults in exchanges, causing groups of bits to be lost or groups of bits from other channels to appear in the middle of an otherwise correct stream of bits.

(iii) The exact timing of the bits advancing and retarding (jitter), leading to the possibility of errors at the receiving terminal.

(iv) Loss of synchronisation between two points in the channel, causing strings of 0s or random bit sequences to replace normal transmission for short periods.

In designing each element of a switched system, it is necessary to relate the various impairments which they introduce into the transmission channel to the effect these have on the user. A large number of experiments must therefore be carried out in order to establish empirical relationships between each type of impairment and user dissatisfaction. This is particularly difficult when a combination of two or more impairments has to be considered. An account of such experiments will not be given here, but may be found in References 21, 22, and 23.

The following are generally accepted units and notation used in dealing with signals and impairments. The attenuation produced by any element of the system, such as a transmission link or part of an exchange, is often measured in decibels (dB). If the power level at the input end of a transmission channel is p_1 watts, and at the output end is p_2 watts (less than p_1), then the channel is said to introduce an attenuation of $10 \log_{10} (p_1/p_2)$ dB. It is often useful to have a standard power level and to relate all power levels to this logarithmically using the dB notation. There are a number of reference levels in use that you may come across in various publications. All these are arbitrarily defined for a particular system or a particular type of measuring instrument. A useful reference level is one milliwatt. A signal power level referred to one milliwatt is written as P dBm, where P is negative for power levels less than a milliwatt, and positive for levels greater than a milliwatt.

Noise is a difficult impairment to measure and describe. What is of interest is not the absolute power of the noise, but a measure of its interference to communication. Different frequencies of noise interfere to different extents, the most disturbing on a 300-3400 Hz telephone channel being around 1000 Hz. What is done, therefore, is to weight the noise power at each frequency according to the weighting function shown in Figure 3.1. This is derived from experiments in which people were asked to judge the

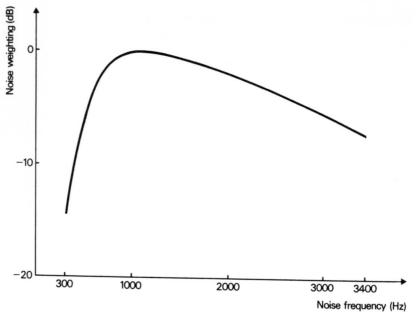

Figure 3.1 Psophometric weighting function

extent to which various tones interfered with conversation. The measurement of noise in terms of its effect on the user is called **psophometry,** and the weighting function in Figure 3.1 is therefore called the **psophometric weighting function.** When a range of frequencies is present in the noise, the use of the psophometric weighting function, together with an integration over the speech band, gives a reasonable measure of the overall interfering effects of the noise. For example, white noise over the 300-3400 Hz band is, when weighted by the function shown in Figure 3.1, 2·4 dB lower than the same basic noise power at 1000 Hz. The unit usually used to express the absolute noise power on a transmission channel is the pW (10^{-12} W). The noise power after psophometric weighting is described as a certain number in pWp. Alternatively, the noise may be expressed in dBm; a psophometrically weighted noise power in dBm is described as a certain number in dBmp.

Devices used for measuring noise (psophometers) contain weighting networks with the characteristic shown in Figure 3.1 so as to give the required weighting. After passing through this weighting network the signals are applied to a suitable power meter, which then indicates the noise level in pWp or dBmp. Because bursts of noise of less than 200 ms have less interfering effect than their power would indicate, a psophometer usually also incorporates a means of reducing its short-burst response. Where the reading given by the psophometer is continuously varying, for a telephone channel a one minute average is usually considered to give a good measure of the interfering effect, since this is comparable with the length of a typical telephone call (about 200 seconds on average).

3.2 TERMINALS

The first element of any telecommunication system, whether switched or non-switched, is the terminal. As mentioned in Chapter 1, the terminal in a switched system generally contains an input transducer and an output transducer. In some cases, however, the terminal may not be a fixed one which can be regarded as a part of the system; the user may be able to connect any one of a number of terminals to the system. In this case it is convenient to regard the electrical interface, between the local line and whatever terminal equipment is connected, as the terminal, even though this interface contains no means of producing message signals itself. An example of such an interface is a data modem connected to a telephone system; this converts binary digital signals from the user's equipment into analogue signals that can be sent over the telephone transmission channel, and vice versa. A comprehensive description of telecommunications terminals cannot be given here. References 16 and 26 describe various terminals which are connected to switched systems and which contain input and output transducers. Reference 27 describes a particular type of data modem which can be connected in place of a telephone terminal. In this section a description of a typical telephone terminal will be given.

The functions that a telephone terminal has to perform are as follows:

(a) Convert the sound of the user's voice into electrical signals which are an analogue of that sound.

(b) Convert similar signals received from the distant terminal back into sound.

(c) Match the outgoing signals to the local line and the incoming signals to the output transducer (the earphone) and make it possible to convey both sets of signals over the one pair of wires forming the local line.

(d) Provide the necessary local signalling facilities described in Chapter 2.

The type of input transducer (microphone) used in most telephones to convert sound into electrical signals is a carbon granule microphone. This does not generate electrical power using the energy in the sound, as some microphones do, but varies the resistance of the electrical path in accordance with the pressure produced on a diaphragm by the sound waves. This is done by applying this pressure to a capsule of carbon granules whose conductive properties vary according to pressure. Power, drawn from the local exchange through the local line, is fed to the microphone so that the changes in its resistance cause the current to vary in sympathy with the sound. An a.c. speech signal is thus superimposed on the d.c. line current. The reason for using a carbon granule microphone is that it is cheap, sensitive and robust. It has non-linear characteristics, with a lower sensitivity to lower sound levels than higher ones. Although this causes distortion of the sound signal, in this application it is generally beneficial because it helps to discriminate between speech and room noise. Other types of microphone have been investigated for possible use, though it is difficult to equal the robustness and cheapness of the carbon microphone. The output transducer (earphone) in most telephones is either some form of moving iron earphone, or, more recently, a rocking-armature earphone, as described in Reference 25.

In all public telephone systems two-wire local lines are used. This means that to deliver the microphone signals to the local line and feed incoming signals from the local line to the earphone requires separation of the two directions of transmission. Unless something is done to prevent it, the signals produced by the microphone, which may be as much as 40 dB higher than the incoming signals, will be fed into the earphone, thus producing an uncomfortably loud sound in the user's ear when he speaks. This unwanted version of the user's own voice is called **sidetone.** In fact, it is not completely unwanted; without any sidetone at all the telephone sounds dead. In practice, however, it is unlikely that too little sidetone will ever be encountered. Most telephone terminals at present in use produce more sidetone than has been shown, by user tests, to be generally preferred.

The way in which sidetone is reduced is shown in Figure 3.2, which is a diagram of the speech circuit of a typical telephone terminal. The ratios of the numbers of turns in the transformer and the values of the components in the balance circuit are chosen to be such that, when the user is speaking, the current flowing through the left-hand part of the circuit is minimal. When the user speaks, the current flowing through the microphone varies, thus delivering a signal to the local line. At the same time a corresponding a.c. voltage appears across the left-hand winding of the transformer and the balance circuit. For complete prevention of sidetone the voltages across these two should be equal and opposite, so that the net voltage across the earphone is zero. In practice this is difficult to achieve because the required

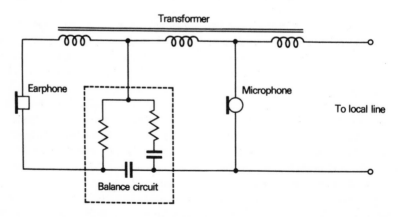

Figure 3.2 Telephone speech circuit

component values depend on the impedance of the local line. Because local lines have different impedances the circuit must be designed to give satisfactory sidetone reduction on the majority of local lines.

An important characteristic of this speech circuit is its sending efficiency, that is the extent to which the signal power from the microphone is delivered to the local line, rather than being wasted by dissipation in the balance circuit. Unfortunately the receiving efficiency, that is the extent to which the incoming signals are delivered to the earphone rather than dissipated in the rest of the circuit, varies inversely with the sending efficiency, so a compromise has to be reached. (This is discussed more fully in Reference 28.) The optimum arrangement is generally when the power from the microphone is shared equally between the local line and the balance circuit.

The parts of the telephone terminal concerned with local signalling were described in Chapter 2. Figure 3.3 shows a complete practical circuit for a dial telephone. Note that the 1·8 μF capacitor, which isolates the bell from the local line as far as d.c. is concerned, is shared between this function and the balance circuit; were it not for this arrangement, two separate 1·8 μF capacitors would be needed. This capacitor, together with the resistors in the circuit, performs a third function; it acts as a spark quench arrangement for

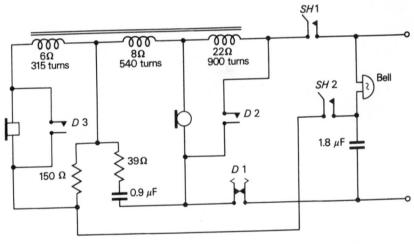

Figure 3.3 Complete telephone circuit, including signalling elements

the pulsing contacts of the dial, $D1$. By minimising sparking when the contacts open, this reduces contact wear and prevents interference between local lines sharing the same cable. Contacts $D2$ and $D3$ in the dial are used to short out the microphone and earphone when the dial is pulled round from its rest position. These contacts disable the speech circuit during dialling so that dial pulses are not affected by the impedance of the microphone and clicking is prevented in the earphone when pulsing occurs.

It is very difficult to make objective statements about the performance of a telephone terminal. On the basis of user tests one can talk about preferred values for various parameters, but to classify things as satisfactory or unsatisfactory requires a definition of these terms that is applicable to the sort of experiments that can be carried out. For example, unsatisfactory might be taken to mean that more than 5 per cent of users would describe the thing in question as poor or bad when given a choice of excellent, good, fair, poor, and bad. Each user's opinion will depend on the characteristics of his speech and hearing and also on other physical factors such as the relative positions of his ear and mouth (which affect the performance of the handset). Some attempts to make quantitative assessments of the performance of telephone terminals are described in References 21 and 23.

To give you some idea of the power levels used for message signals in a telephone system some examples will be given. The following figures are approximate. With the microphone of the handset in front of the mouth, a typical user's speech causes the microphone to give an a.c. signal power of the order of 0·2 mW (− 7 dBm). Assuming that this is shared equally between the balance circuit and the local line, a signal power of the order of 0·1 mW (− 10 dBm) is delivered to the local line. The attenuation produced by the overall transmission channel between the two terminals (as measured at 1 000 Hz) may be anything from about 10 dB to 40 dB, so the signal level at the other terminal may vary between − 20 dBm and − 50 dBm. Assuming that half the received power is delivered to the earphone and half is dissipated in the rest of the speech circuit, the power applied to the earphone can vary between − 23 dBm (5 μW) and − 53 dBm (5 nW). Only a few per cent of this power is converted into sound power by the earphone, but a high proportion of the sound power is delivered to the inner ear, provided that the handset is placed firmly against the outer ear. In the worst case, a majority of users are able to hear incoming speech adequately, provided that the room they are in is reasonably quiet and the transmission channel is moderately free of noise. In the best case, some users find the sound level too high, so they hold the earphone slightly away from the ear. The preferred sound level is produced by a signal of about − 26 dBm (2·5 μW) applied to a modern earphone.

3.3 TRANSMISSION SYSTEMS

The simplest method of providing a transmission link is using a single pair of wires. This involves a high cost per kilometre as compared with other types of transmission link, but the cost of the equipment needed at the ends of the link is comparatively low. It is usually the cheapest method where the length of the link is only a few kilometres or where there are only a few links on a particular route. It is widely used for local lines and trunks of a few kilometres length. Where more than about 20 transmission links are on the same route and the route is longer than a few kilometres, it is cheaper to use a method of transmission which involves multiplexing a number of channels to form a channel of higher transmission capacity. The cost per kilometre is then lower, but the equipment at the ends of the link is more expensive. Provided that the distance covered is more than a few kilometres, the extra end costs are more than covered by the savings in line costs. In general, the more channels that are carried by a single transmission system, the lower the cost per channel. As mentioned in Reference 48, the cost of a system with N speech channels is roughly proportional to $N^{1/3}$, so the cost per channel is proportional to $N^{-2/3}$.

In this section we are going to look first at the basic properties of wire pairs, which are widely used as transmission media. We shall then look at various methods of using wires to provide transmission links. A brief description of line-of-sight microwave radio systems is included at the end of the section.

3.3.1 The properties of wire pairs

The characteristics of a pair of wires are determined by the primary constants: R (the resistance per unit length), L (the inductance per unit length), G (the conductance between the pair per unit length), and C (the capacitance between the pair per unit length). These are measured as a.c. parameters and are, to various extents, functions of the frequency. The most important frequency dependence is the variation of R as the square root of the frequency at high frequencies, due to what is known as the skin effect. The transmission properties of a wire pair may be represented by the secondary constants, Z_0 (the characteristic impedance) and γ (the propagation constant). γ is normally separated into its real and imaginary parts, α and $j\beta$, where α is the attenuation index and β is the phase shift index. As shown in Reference 6, Z_0 and γ are related to R, L, G, and C by the expressions:

$$Z = [(R + j\omega L)/(G + j\omega C)]^{1/2} \tag{3.1}$$

$$\gamma = [(R + j\omega L)(G + j\omega C)]^{1/2} \tag{3.2}$$

where ω is the angular frequency ($=2\pi f$). α, the real part of γ, can be used to derive the attenuation per unit length by the relation:

$$\text{Attenuation in dB per unit length} = 20 \log_{10}(e^\alpha) = 8 \cdot 686\alpha \tag{3.3}$$

An analysis of the behaviour of Equation (3.2) over various frequency ranges for practical values of the primary constants shows that there may be four regions as follows:

(1) Where $\omega \ll R/L$ and $\omega \ll G/C$, α is independent of frequency and given by $\alpha \simeq (RG)^{1/2}$.

73

(2) Where $G/C \ll \omega \ll R/L$, α varies as the square root of the frequency and is given by $\alpha \simeq (\omega CR/2)^{1/2}$.

(3) Where $\omega \gg R/L$ and $\omega \gg G/C$, and where R is more or less independent of frequency, α is given by $a \simeq (R/2)(C/L)^{1/2}$.

(4) Where $\omega \gg R/L$ and $\omega \gg G/C$ and R is determined largely by the skin effect which causes R to be proportional to $\omega^{1/2}$, α is proportional to $\omega^{1/2}$.

In a particular cable two of these regions may merge. For example, regions (2) and (4) may come together to form an almost continuous $\omega^{1/2}$ region.

The earliest form of wire pair consisted of open (uninsulated) wires strung between ceramic insulators fixed on poles. These are rarely used nowadays. Most wire pairs are now in the form of cables. These contain a number of pairs and are normally buried in the ground in ceramic, metal or asbestos ducts. This gives them physical protection and, more importantly, makes it possible to bury new cables by pulling them into a duct which already contains some cables; only when a duct becomes full does digging become necessary to bury a new duct. Where regular increases in the number of cables are unlikely, cables may be buried without a duct. In rural areas this can be done using a mole plough which ploughs the cable straight into the ground as it is pulled along. Where poles exist by the side of a road, cables can be strung between the poles, though this looks unsightly.

There are two forms of widely used wire pair. The first is the twisted pair, consisting of copper or aluminium conductors insulated with paper or plastic and contained in a lead or plastic outer sheath. Up to about 4000 pairs are used in modern cables and the larger cables are normally pressurised, that is the air inside them is maintained above atmospheric pressure so that water does not seep in if a leak develops; leaks can be detected by a fall in pressure. The second form of wire pair is the coaxial tube, consisting of a centre copper wire supported in a thin copper tube by means of plastic discs through which the wire is threaded. The discs are spaced, typically, 3 cm apart. The tubes are bound with mild steel tapes to give added strength and electromagnetic screening, and are insulated with paper. Up to about 20 tubes can be contained in a cable, protected by an outer lead sheath which is usually coated in polyethylene. Coaxial cables are generally pressurised.

Twisted pair cables have copper or aluminium conductors which range from 0·32 mm to 1·27 mm diameter. Aluminium cables are made with conductor diameters 4/3 times their copper equivalents because of the higher resistivity of aluminium. This means that, although aluminium is cheaper than copper, some of the savings are lost through the extra duct space used up by the thicker cables. Typical values of the primary constants for a 0·63 mm diameter copper twisted pair are: $R \simeq 100 \ \Omega \ \text{km}^{-1}$, $L \simeq 1 \ \text{mH km}^{-1}$, $G \simeq 10^{-5} \ \text{S km}^{-1}$, and $C \simeq 0.05 \ \mu\text{F km}^{-1}$. Thus, for most practical purposes, twisted pairs are used in region (2), the first $\omega^{1/2}$ region. The attenuation produced by a twisted pair therefore increases with frequency and, even using the thicker gauge conductors, exceeds 5 dB km^{-1} at around 500 kHz.

However, it is not attenuation that places an upper practical limit on the operating frequencies of twisted pairs. The most important consideration is crosstalk, caused by capacitative coupling between pairs in the same cable. The effects of this can be reduced by complicated balancing networks of resistors and capacitors at each end of the cable. Nevertheless, crosstalk imposes a practical upper limit of 500 kHz on operating frequencies.

Coaxial cables were first introduced in 1937 in order to overcome the crosstalk limitations of twisted pairs for high capacity transmission systems. At frequencies of the order of 1 MHz, where crosstalk would otherwise become a problem, the skin effect causes the signal currents to be confined to the inner surface of the outer conductor, whereas outside interference will cause currents to flow in the outer surface of the outer conductor. Because of the skin effect, attenuation is proportional to $\omega^{1/2}$ and, at a given frequency, is inversely proportional to the conductor diameters (for a fixed diameter ratio). As described in Reference 13, for a given diameter centre conductor, α is minimised by making the internal diameter of the outer conductor 3·6 times the diameter of the inner one. Coaxial cables are referred to according to these two diameters. Two standard tube sizes recommended by the CCITT are 2·6/9·5 mm and 1·2/4·4 mm, which both have this diameter ratio of 3·6. Coaxial tubes produce much less attenuation per unit length than twisted pairs made out of the same amount of copper.

3.3.2 Audio transmission links

An audio transmission link is a single transmission link providing just one speech channel. The simplest example of such a link is a local line. An audio link may be either a pair of wires, like a local line, or two pairs of wires, one for each direction of transmission. A link using separate channels for the two directions is therefore referred to as a **four-wire** link. A two-wire channel may be converted to a four-wire channel either to introduce amplification or to facilitate multiplexing of several channels, as will be described later.

In order to convert a two-wire channel into a four-wire one, a device called a **hybrid** is used. This normally consists of two transformers, arranged as in Figure 3.4 (Figure 3.5 shows the symbolic representation of a hybrid.) The signal power arriving at the hybrid from the two-wire channel is shared between the two transformers. The signals thus pass to the first amplifier on the go channel. Signals also pass down the return channel, but cannot pass in the wrong direction through the first amplifier. Signals arriving from the return channel are shared between the balance impedance and the two-wire channel. Provided that the balance impedance exactly matches the impedance of the two-wire channel, and the windings in the transformers are well matched with one another, then the signals in the windings W_1 and W_2 cancel each other, so no signals are fed back down the go channel from the return channel. In practice it is not possible to exactly match the balance impedance to the channel impedance and some power therefore passes from

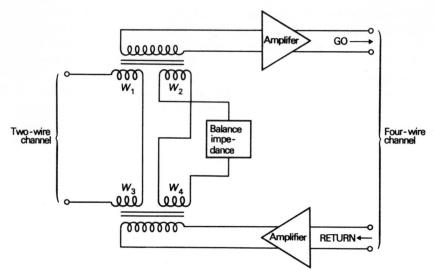

Figure 3.4 Hybrid circuit

the output of the return amplifier to the input of the go amplifier. If this happens at both ends of a four-wire link, a transmission loop is created. Therefore if too much amplification is provided in the go and return paths of a link such as the one shown in Figure 3.6, oscillation (or singing as it is sometimes called) will occur. Even if the gain round the loop only approaches unity, this produces selective amplification of certain frequencies for which the net phase shift round the loop is zero. This causes the link to sound hollow to the users. It is found that in practice the loop loss must be at least 6 dB to prevent this effect. This is known as the **singing margin.** Taking these effects into account, and making allowance for the fact that the actual gain of an amplifier may vary from its design value, it is found that the maximum amplification that can be put into the channels of a practical four-wire link is such that the loss of the overall link, between the two-wire points, can be reduced to no less than about 3 dB. This is discussed more fully in Reference 13.

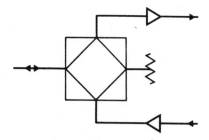

Figure 3.5 Symbolic representation of a hybrid

The arrangement shown in Figure 3.6 can be used to provide transmission links of several tens of kilometres length using two wire pairs for each link. Repeaters, containing an amplifier for each direction of transmission, are inserted at regular intervals to maintain the signal level at a high level relative to noise, in particular to noise caused by crosstalk. For links of 30 km or less it is not essential to use a four-wire channel. A single pair of wires can be used as a transmission link and, if necessary, a certain amount of amplification can be introduced by one of two methods, as described below.

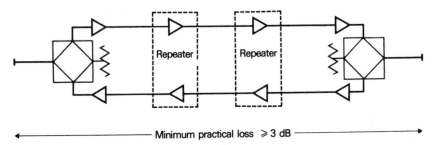

Figure 3.6 Four-wire audio link with repeaters. Each line represents a wire pair

Figure 3.7 illustrates four common types of two-wire link. The first is simply a pair of wires without any amplification. Links of up to about 20 km can be of this form. For longer links, or links on which thinner gauge wires are to be used, it may be necessary to introduce amplification. The first method of doing this is with a **hybrid repeater.** This converts the two-wire link to four-wire form and back again using two hybrids and introduces amplification into the two channels in between the hybrids. To prevent singing, the balance impedances must be very carefully matched to the line impedances. The maximum reduction of loss per repeater achievable with hybrid repeaters is about 12 dB. A hybrid repeater may be attached to one end of the link, as shown in the second arrangement in Figure 3.7, but is best placed near the centre, as shown in the third arrangement. With a repeater near the centre it is easier to achieve reliable balance impedance matching.

The fourth arrangement shown in Figure 3.7 uses the second method of introducing amplification. This consists of a special type of two-wire repeater containing a **negative impedance convertor,** or **n.i.c.** for short. An n.i.c. is a one-port (two-terminal) device which is such that the power of a signal applied to its terminals is increased; power flows out of the device, rather than into it, so it behaves as a negative load or impedance. It is basically an amplifier with positive feedback from its output into its input, the amount of feedback being carefully controlled to prevent oscillation. As with a hybrid repeater, an n.i.c. repeater requires careful impedance matching between the line and a balance impedance in the n.i.c. To ensure stability against oscillation, normally only one n.i.c. repeater is used per link, and this is placed near the centre of the link. The maximum reduction of loss that can be obtained using an n.i.c. repeater is about 10 dB. Reference 55 describes some practical aspects of n.i.c. repeaters.

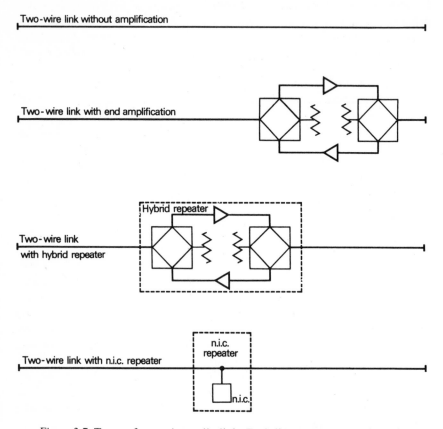

Figure 3.7 Types of two-wire audio link. Each line represents a wire pair

The wire pairs used for audio transmission links are normally twisted pairs. Because α is in the first $\omega^{1/2}$ region for the 300-3400 Hz speech band there would be considerable attenuation at the high frequency end of the band on links of more than a few kilometres length. Also, the variation of attenuation with frequency would cause impairment of the speech signals. However, Equation (3.2) can be used to show that in the special case where $GL = RC$, α becomes independent of frequency for all frequencies, and equal to the low frequency value, $(RG)^{1/2}$. In practical cables, L is very much lower than the value needed to achieve this condition, so if L can be increased to the value RC/G, the transmission characteristics of a wire pair can be greatly improved. Increasing L for this purpose is called **loading** the line. In fact, it is very difficult to load a line. It can be done by wrapping permalloy tape or wire round each conductor, but this is prohibitively expensive.

It is found that, in practice, a significant improvement in the characteristics of a wire pair can be achieved by adding only a proportion of the inductance required to make L equal to RC/G. Also, it is found that the inductance does not have to be uniformly distributed, provided that only a limited frequency range is to be used; the inductance can be added in lumps at regular intervals: this is called **lumped loading.** It has a similar effect to continuous loading up to a certain cut-off frequency, above which the attenuation rises rapidly. The cut-off frequency is lower the further apart are the loading inductors, or **loading coils** as they are often called. Because the cost of loading a cable depends mainly on how many loading coils are inserted, the most economic arrangement is to have the distance between the loading points such that the cut-off frequency is just above the high frequency end of the band to be used. In Britain and North America, a widely used arrangement for 300-3400 Hz links using twisted pairs is one using 88 mH loading coils spaced 1·83 km (=2000 yards) apart. Loading is used on many audio trunks, but it is not normally economic for local lines.

3.3.3 Transmission systems using f.d.m.

Most modern f.d.m. systems use coaxial tubes to carry the signals. In most inland systems one tube is used for each direction of transmission, so each transmission system uses two coaxial tubes. A cable, which may contain as many as 20 tubes, can therefore serve a number of transmission systems. For undersea cables it is more economic to use a single coaxial tube and to share this between both directions of transmission. In this case one half of the available bandwidth is used for one direction and the other half for the other. This makes the repeaters more complex because the two directions of transmission have to be separated by means of filters before amplification can be carried out. Twisted pairs can also be used for f.d.m. transmission systems, but crosstalk limits the maximum operating frequency to about 500 kHz. Using one pair for each direction of transmission, this provides up to 120 channels for every two pairs. A further limitation is that it is necessary to use twisted pairs in separate cables for the two directions of transmission because of crosstalk. Transmission systems using f.d.m. signals over twisted pairs are now obsolescent.

It is assumed here that you are familiar with the basic principles of f.d.m. using single sideband (s.s.b.) modulation. A typical practical multiplexing scheme will now be described. Each channel, if not already in four-wire form, is split into separate go and return channels by means of a hybrid. In the first stage of multiplexing, each of 12 go channels is used to modulate a different carrier, with s.s.b. modulation, and the resulting signals are combined. As part of the modulation process the signals corresponding to each channel are limited to a bandwidth of 3·1 kHz by means of filters. The carriers are in steps of 4 kHz from 64 kHz to 108 kHz. The gap of 0·9 kHz thus left between adjacent channels allows simpler filters to be used. The lower sidebands are selected during the modulation process, so the resulting assembly of 12 channels, which is called a **group,** is in the band 60-108 kHz.

In the second stage of multiplexing, five groups are used to modulate carriers in steps of 48 kHz between 420 kHz and 612 kHz. Again, the lower sidebands are selected and the resulting assembly of 60 channels, known as a **supergroup,** occupies the band 312-552 kHz. In the third stage of multiplexing, a number of supergroups are used to modulate carriers in steps of 248 kHz from 1116 kHz upwards. The lower sidebands are again selected, so the result is a band from 564 kHz upwards. In order to utilise the part of the frequency spectrum below 564 kHz, another supergroup is used to modulate a 612 kHz carrier which, after selection of the lower sideband, produces a supergroup brought down to a frequency band 60-300 kHz. The gap between 300 kHz and 564 kHz is then filled with a further supergroup, in its basic form, that is 312-552 kHz. Note that there is a 12 kHz gap between the supergroups 60-300 kHz, 312-552 kHz, and 564-804 kHz, but an 8 kHz gap between the other supergroups. There are various names given to combinations of different numbers of supergroups, but the most important is the **hypergroup.** This is usually a combination of 15 or 16 supergroups, containing 900 or 960 channels.

Once all the supergroups have been assembled into a single band, this is delivered to the coaxial tube. At the far end a demultiplexing process, which is the exact inverse of the multiplexing process, is carried out: the band is broken down into supergroups, which are broken down into groups, which are broken down into channels. Where a number of transmission links have to be made up out of channels from f.d.m. systems joined end to end, a complete group or supergroup can be filtered from one system and patched through to another. This saves the multiplexing equipment that would be needed to patch through individual channels. It also removes the possibility of extra noise being introduced by the demodulation and remodulation processes.

In order to use coaxial cable over a particular frequency band, it is necessary to ensure that, first, the repeaters, which are inserted at regular intervals to maintain the signal level at a high level relative to noise, can handle the required frequency band. Secondly, it is necessary to ensure that the repeater spacing is small enough for the attenuation produced by the cable at the maximum frequency to be compensated for by the repeaters. For example, a 2700-channel system has a maximum signal frequency of about 12 MHz. Using 2·6/9·5 mm coaxial tubes the attenuation at 12 MHz is about 8 dB km^{-1}. A typical 12 MHz repeater is only able to satisfactorily provide about 36 dB amplification, so a repeater spacing of about 4·5 km must be used. The main factor limiting the gain that can be provided by amplifiers in repeaters is harmonic distortion. A non-linear amplifier which amplifies two sinusoidal signals of frequency ω_1 and ω_2 will produce additional signals with frequencies $2\omega_1$, $2\omega_2$, $(\omega_1 + \omega_2)$, $(\omega_1 - \omega_2)$, $(2\omega_1 + \omega_2)$, $(2\omega_1 - \omega_2)$, and so on. These extra components are called **intermodulation products.** The intermodulation products arising from non-linear amplification of a complex signal will cover a wide range of frequencies, and the collective effect of those which are within the transmission band will be to add noise to the speech signals in the individual channels. It is usually the practical difficulties of designing amplifiers which are sufficiently linear to ensure an acceptably low level of intermodulation noise which limit the gain that can be provided at each repeater.

In order to keep the attenuation of a complete transmission system constant, some repeaters contain amplifiers whose gain is automatically varied to compensate for changes in attenuation due to changes in the temperature of the cable. Certain types of repeater achieve this by means of a temperature sensing device. This only approximately compensates for variations in attenuation. In order to achieve complete correction, a single frequency pilot signal is sent at a fixed level by the multiplexing equipment and a small number of repeaters monitor the level of this signal and adjust their gain so as to exactly cancel all remaining variations in attenuation. The repeaters which carry out this correction are more complex and costly than other repeaters; this is why only a small number of these are used in each transmission system.

To compensate for the variation of attenuation with frequency in the cable, the gain/frequency characteristics of repeaters are made complementary to the attenuation/frequency characteristics of each section of cable. The process of compensating for the attenuation/frequency characteristics of the cable is known as **equalisation.** In modern repeaters the exact gain/frequency characteristic is controlled by a device which monitors several pilot signals at different frequencies throughout the transmission band, so that variations in the attenuation/frequency characteristics are compensated for.

3.3.4 Transmission systems using t.d.m.

It is assumed here that you are familiar with the principles of pulse code modulation (p.c.m.), that is with the way an analogue signal can be converted to a binary digital signal by a process of sampling and encoding. According to Nyquist's sampling theorem, the sampling rate in the p.c.m. process must be at least twice the maximum frequency present in the analogue signal. In most practical systems a sampling rate of 8000 samples per second is used. Because of the wide range of signal levels that are found on telephone channels, user assessments of p.c.m. channels have shown that it is better to encode the samples using digital codes which represent voltage levels that are not evenly spaced in the signal voltage range. The levels are closer together for lower voltages so that, for different average signal levels, the voltages changes between one level and the next bear a roughly constant relationship to the signal level. This means that the level of quantisation noise relative to the signal level is roughly independent of the signal level. The encoding process is said to be done using a **companding law.** Details of the companding laws used in various systems are given in References 8, 9 and 48.

For a single p.c.m. link, acceptable reproduction of speech can be obtained with 7 or even 6 bits for the digital code representing each sample. However, one of the standard systems recommended by the CCITT uses an 8-bit code. This ensures that acceptable transmission quality is obtained when several p.c.m. links are used end to end with the signals being converted back to analogue form in between the separate links. The channels corresponding to each speech channel after digital encoding have a bit-rate of $8 \times 8000 = 64000$ bits s^{-1}.

The multiplexing technique generally used for binary digital channels is that of time division multiplexing (t.d.m.). The duration of the bits in each of n channels is reduced by a factor n, so that a block of 8 bits from a given channel can be transmitted in $(1/8000n)$ second, instead of $(1/8000)$ second. A $(1/8000n)$ second time-slot is allocated to each of the n channels in rotation, so that 8 bits from one channel follow 8 bits from another, and so on. The n channels are thus multiplexed to form one channel with a bit-rate of $(n \times 64000)$ bits s^{-1}.

A multiplexing hierarchy, similar to the group/supergroup/hypergroup hierarchy for f.d.m., has been defined for t.d.m. by the CCITT. The hierarchy is shown in Table 3.1. A 30-channel t.d.m. group actually consists of 32 channels, but two of these are not used as active speech channels. One is used for control of synchronisation of the equipment at the two ends of a link and the other is reserved for use as an inter-exchange signalling channel. A t.d.m. supergroup incorporates four spare channels for various control purposes, and a hypergroup incorporates two spare groups.

Table 3.1. *Multiplexing hierarchy for t.d.m. systems*

Unit	Number of 64 kbits s^{-1} channels used for speech	Bit-rate
One channel	1	64 kbits s^{-1}
Group (30 channels)	30	2·048 Mbits s^{-1}
Supergroup (4 groups)	120	8·448 Mbits s^{-1}
Hypergroup (16 supergroups)	1920	139·264 Mbits s^{-1}

Two twisted wire pairs (one for each direction of transmission) may be used to carry one t.d.m. group. The crosstalk considerations, which restrict the use of twisted pairs to about 500 kHz with analogue signals, do not affect digital signals up to about 2 Mbits s^{-1}, because the difference between a binary 0 and a binary 1 can be discerned with a very low probability of error in the presence of a considerable amount of noise. However, signals at a bit-rate of 2 Mbits s^{-1} are attenuated to such an extent that it is necessary to have repeaters as close together as about 2 km. Systems of this type, providing 30 active speech channels and using 2 channels for synchronisation and signalling, are often referred to as $30+2$ channel systems. Their use has considerable economic advantages if twisted pairs are already in existence and being used as audio transmission links. In this case two of these pairs can be taken over for use on a $30+2$ system, so that, counting the loss of the two audio links, there is a net gain of 28 channels for the price of the multiplexing equipment and repeaters. The lumped loading has to be removed from the twisted pairs and the repeaters are designed to fit in place of the loading coils, the repeater spacing thus becoming, typically, 1·83 km.

For high capacity t.d.m. systems coaxial tubes are used to carry the signals. The repeater spacings required are comparable with those for f.d.m. systems with similar numbers of speech channels. For example, to carry a t.d.m. hypergroup (1920 active channels) using 2·6/9·5 mm coaxial tubes (one for each direction of transmission) requires a repeater spacing of about 4·5 km, the same as for a 12 MHz (2700 channel) f.d.m. system.

In order to take advantage of the fact that a binary digital signal can be separated from a considerable amount of noise by deciding whether the signal level represents a 0 or a 1, a selection process of this sort can be carried out in the repeaters. Thus, rather than simply amplifying the incoming signals, a digital repeater may incorporate a device which identifies signals as 0s or 1s and then completely regenerates them as effectively noise-free signals. A repeater working in this way is called a **regenerative repeater.** In order to be able to identify each incoming pulse, a regenerative repeater contains a clock which is synchronised with the incoming pulse stream. In a simple system, an internal clock in the repeater contains a resonant circuit which is stimulated by the incoming pulses so that it oscillates in step with the 1 pulses, and continues oscillating on its own during the 0 pulses (which are really no pulses). Figure 3.8 shows the basic functional blocks of a regenerative repeater. The part within the dotted line is peculiar to a regenerative repeater, whereas the part outside is basically similar to an ordinary repeater. Regenerative repeaters are therefore more costly than ordinary ones, so most long distance t.d.m. transmission systems use a mixture of ordinary and regenerative repeaters for economy; it is not necessary to regenerate the pulses at every repeater when coaxial tubes are used.

A difficulty that arises from the use of an internal clock is that, when a speech channel is not in use, a string of eight 0s appear in its time-slot. At night, when very few telephone calls are made, several adjacent time-slots may be full of 0s. During a long string of 0s the internal clock is likely to lose synchronism. To prevent this happening, every bit in each time-slot can be inverted in the multiplexing equipment. So, 00000000 becomes 01010101. At the far end the bits are reinverted to restore them to their correct values. This arrangement ensures that there is a very low probability of there being more than a few 0s in succession. Other methods of preventing loss of synchronism are used, but this is the simplest. It is normally referred to as **alternate digit inversion.**

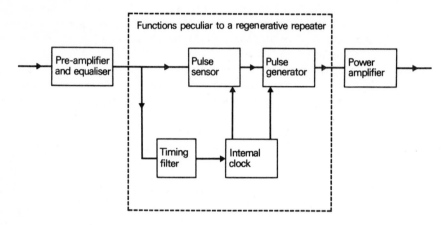

Figure 3.8 Functional blocks of a regenerative repeater

Another difficulty that arises in the transmission of digital signals is that, in general, the number of 0s and the number of 1s are not equal. This means that there is, on average, a d.c. component to the signals, if 0 is treated as $-V$ volts and 1 is $+V$ volts. In order to make it possible to pass the signals through a transformer in the repeaters, the signals must be converted to a form in which there is zero d.c. content on average. This can be done by using ternary (three-level) signals, with values $-V$ volts, 0 volts, and $+V$ volts. A binary 0 is then represented by 0 volts, but a binary 1 is represented by either $-V$ volts or $+V$ volts. These two signals for a 1 are used in alternation, so that the resulting signals have an average voltage level of zero; there is thus no d.c. component. This technique is referred to as **bipolar transmission.** It is carried out after alternate digit inversion. There are other, more sophisticated methods for bipolar representation of binary digital signals, as described in Reference 48.

In passing through the cable the signals are changed from the form in which they are sent to somewhat rounded pulses. This is due to the fact that the higher Fourier components of the signal are attenuated more than the fundamental frequency during transmission. The emerging waveform thus has a stronger sinusoid component of the fundamental frequency, having lost the sharp edges which are defined by the higher Fourier components. This is illustrated in Figure 3.9. In many systems the signals delivered to the line therefore contain pulses (both $-V$ volts and $+V$ volts) which are only half the nominal width of a binary signal. The pulses are then said to have a **50 per cent duty cycle.** As illustrated in Figure 3.9, this makes the received pulses more easily distinguishable.

To summarise what has been said about the use of t.d.m. for speech signals, the sending process can be split into five parts: digital encoding using p.c.m., multiplexing, alternate digit inversion (or equivalent), conversion to bipolar, and reduction of the duty cycle. The receiving process consists of the inverse of each of these in the reverse order.

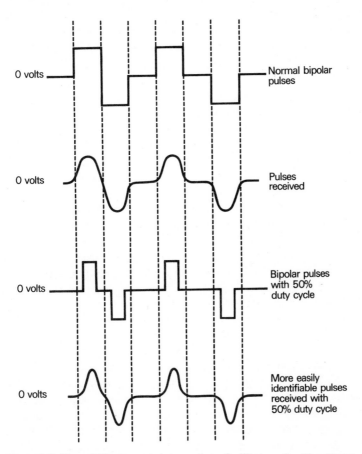

Figure 3.9 Use of 50 per cent duty cycle to facilitate pulse identification

3.3.5 Microwave radio systems

As a last example of a transmission system, a brief description of a line-of-sight microwave radio system will now be given. The frequencies widely used for line-of-sight microwave radio are in the range 1-12 GHz. (Experiments in the use of higher frequencies have shown that these are subject to severe fading in poor weather conditions, but can be used under certain circumstances.) The radio waves are beamed between antennae, which are either parabolic or horn shaped. These are mounted on the tops of towers 100 metres or more high, so as to provide adequate clearance of trees and buildings, thus preventing interference due to signal reflection and diffraction. The towers are spaced 40 to 50 km apart. Each system consists of two terminal stations and a number of intermediate repeater stations. Signals are attenuated by roughly 140 dB over 40 km (reduced to 60 dB by antenna gain), so this attenuation has to be compensated for by each repeater.

Between any two antennae a number of carriers are used. Each antenna can be used both to transmit and receive, as long as the carrier frequencies for the two directions of transmission are well separated. Carriers are spaced about 30 to 40 MHz apart, and are grouped in bands containing up to 16 carriers each. The bands are used in pairs, one for each direction of transmission, and the two bands in a pair are separated by about 70 MHz or more. Each pair of bands is referred to by the centre frequency in the gap between the bands. The widely used centre frequencies, as recommended by the CCIR, are: 1·8 GHz, 2·1 GHz, 4·0 GHz, 6·175 GHz, 6·76 GHz, and 11·19 GHz. The pairs of bands at the lower frequencies contain fewer carriers than those at the higher frequencies. In all, about 100 carriers are used, 50 in each direction.

The carriers are modulated by a signal derived from multiplexing a number of speech channels, using f.d.m. or t.d.m. The signals delivered to the transmitter are therefore similar to the signals delivered to the coaxial tube in a cable transmission system. In many microwave systems the modulating signal consists of 16 or 30 f.d.m. supergroups. Television channels can also be used, in place of the 30 supergroups, and most microwave systems carry a mixture of telephone and television channels. Most systems at present use frequency modulation of the carriers, although in this application this makes less efficient use of the frequency spectrum than s.s.b. modulation. Frequency modulation is used because it is easier to make repeaters with the required degree of linearity than with s.s.b. modulation.

Horn-shaped antennae, although more expensive than parabolic antennae, are widely used because they are able to handle several bands at once, making the cost per band less than with individual parabolic antennae for each band. A typical horn antenna can handle three pairs of bands, that is about 30 carriers in each direction. With two antennae per tower it is thus possible to cover the whole of the useful frequency spectrum between that tower and another tower. Each tower may have antennae pointing in several different directions to other towers and between any two towers a total of over 100000 telephone channels can be provided.

Because changes in atmospheric conditions cause occasional fading of the signals, it is necessary to reserve a number of carriers to act as protection channels. These take over from another carrier when severe fading occurs around a particular carrier frequency; fading tends to occur in only one part of the frequency spectrum at a time. Protection channels also provide protection against equipment failure, since each carrier is associated with substantially individual equipment. Change over from the normal channel to a protection channel between any two towers is initiated automatically whenever the level of a single-frequency pilot signal falls below a certain level. The coordination of a change over between two towers is achieved by means of signals sent over a narrow-band signalling channel in between the main carrier frequencies. (The channel associated with each carrier is one containing up to 1800 speech channels, or more in some systems, so without a protection channel this number of telephone calls would be interrupted.) Normal variations of attenuation are compensated for by automatic gain control in the repeaters.

A more detailed description of microwave radio systems, including the design of the transmitters, receivers and antennae, is given in References 45, 48 and 49.

3.4 EXCHANGES

We now come to the third element of a switched telecommunication system—the exchange. In this book, exchange has been taken to mean automatic exchange, that is an exchange in which all switching operations are carried out without the assistance of a human operator. There are still a number of manual telephone exchanges throughout the world, that is exchanges which rely entirely on operators for the connection of calls, but these are not dealt with here. Reference 7 describes manual exchanges in detail. Of course, operators still play an important part in the operation of automatic systems. They are needed to help users who have difficulties with calls, or need some special service such as transferred charge.

3.4.1 The basic structure of circuit switched exchanges

From what you have read in Chapter 2, you should appreciate that circuit switched exchanges are basically switching mechanisms which can be remotely controlled by either a user or another exchange. The only difference between these two cases is in the type of signalling used to exercise this remote control (local signalling in one case, inter-exchange in the other) and in the exact meaning attached to the address information sent to the exchange. For the moment we shall ignore any difference between local lines and trunks, and simply consider these as transmission links which must be interconnected by the exchange. For convenience, local lines and trunks will be referred to collectively as **lines.**

It should be noted that a line may be used for calls set up in both directions, as is the case for most local lines and some trunks, or one direction only; the use of trunks for calls set up in only one direction simplifies inter-exchange signalling procedures as mentioned in Chapter 2. We shall therefore be referring to incoming lines, outgoing lines, and bothway lines. This should not be confused with the incoming and outgoing directions of transmission. The two directions of transmission on four-wire links will be referred to as the incoming channel and outgoing channel.

In Chapter 2 it was mentioned that the instructions supplied by the user or the other exchange are executed by a control. We shall assume for the moment that there are many control devices in an exchange, and that one is associated with each line. The basic structure of an exchange can thus be viewed as shown in Figure 3.10. Each line has a signalling unit associated with it. This sends and receives the local or inter-exchange signals. It passes the received signals to its associated control, which can do two things: first, operate switches in the switching equipment, so as to connect lines to one another and, secondly, command the signalling unit to send signals back down the line to the user or other exchange.

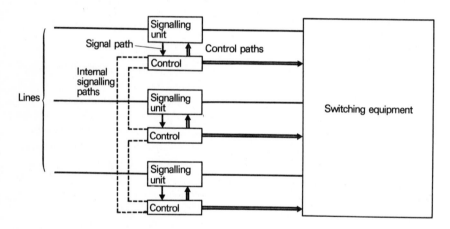

Figure 3.10 Structure of an exchange with individual controls for each line. Only three lines are shown

You will remember from Chapter 2 that, in setting up a call, the control establishes whether a line is free or busy. To do this there must be a means by which one control can interrogate another. Also, when one control has set up a connection through the switching equipment, it must be able to inform the other control of this fact, and get the other control to send the necessary signals over its line (for example, an alert-terminal signal). The means by which one control interacts with another is called **internal signalling.** The internal signalling paths are represented by dotted lines in Figure 3.10. In practice they may consist of circuits switched through the switching equipment along with the inter-line connections. It is also possible to use the inter-line connection itself for internal signalling.

The function of the switching equipment is to interconnect any two lines, along with any internal signalling wires between their controls. The most obvious way of connecting any two lines together is using a matrix of switches, as shown in Figure 3.11. The switches in this kind of matrix are known as **crosspoints.** To connect an inlet of the matrix (a row) to an outlet (a column), the crosspoint is operated where the inlet and outlet wires cross one another. In practice, there would be several such matrices working in parallel, one for each wire. During the rest of this chapter, however, only one wire will be shown on diagrams of switching equipment. This will represent all the wires to be switched for each line. Note that, in Figure 3.11, it has been assumed that calls are set up in one direction only through the matrix. Bothway lines are therefore connected to an inlet and an outlet, whereas incoming and outgoing lines are connected only to the appropriate side of the matrix.

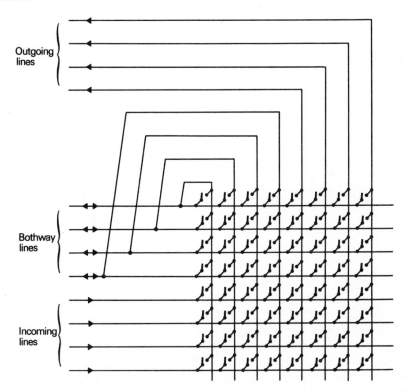

Figure 3.11 Use of a matrix of switches to interconnect any two lines. Only one of the two wires of each line is shown

The picture of an exchange described so far is of a highly inefficient one which would be very uneconomic in practice. There are two areas in which economies are possible. First, the control equipment, which is very complex and expensive, lies idle whenever a line is not in use. Much of its circuitry is only used for a fraction of a second during the setting up of a call. The exchange can be made much more efficient if only a few pieces of control equipment are used, and these are shared between lines by being associated with a call only during those phases of the call when actually needed. The second area of inefficiency in the exchange as described so far is in the switching equipment. To make it possible to interconnect N bothway lines, or N incoming and N outgoing lines, N^2 crosspoints have been used for each wire. Yet, of the N^2 crosspoints per wire, only a very few are used, even when every line is in use at once. For example, if all the lines are bothway and there are 1000 of them, the maximum number of conversations possible is 500. So, out of 1000000 crosspoints, only 500 would be used at once, that is, 0·05 per cent. In practice it would be very unlikely that all lines would be in use at once. There might be, typically, up to 100 simultaneous calls, making the crosspoint utilisation only 0·01 per cent. This utilisation can be considerably improved on, as you will see in a moment.

3.4.2 Efficient switching equipment structures

As an example, we shall consider how the switching equipment in an exchange dealing with 1000 incoming and 1000 outgoing lines (or 1000 bothway lines) can be more efficiently structured than a 1000 × 1000 matrix of crosspoints. A switching matrix will be represented as in Figure 3.12,

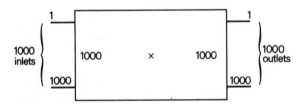

Figure 3.12 Symbolic representation of a 1 000 × 1 000 switching matrix

which shows the 1000 × 1000 matrix. Suppose that the maximum number of simultaneous calls that is likely to occur is 100. The 1000 × 1000 matrix can be replaced by two matrices, one 1000 × 100 (that is 1000 inlets and 100 outlets), and one 100 × 1000, as shown in Figure 3.13. This switching arrangement, or **switching network** as it is more often called, is said to consist of two **switching stages,** connected together by 100 **links.** For convenience, only the first and hundredth links have been shown in Figure 3.13. This switching network can connect up to 100 calls at once, one on each link. However, if there are more than 100 calls to be connected, then blocking will occur. Blocking within the switching equipment is normally referred to as **internal blocking,** to distinguish it from blocking which occurs when all the trunks on a trunk route are in use. We shall suppose, for this example, that the average traffic per incoming line is 0·05 erlang, so that the total traffic to be handled by the switching network is 50 erlangs. In this case the probability of there being more than 100 simultaneous calls at a particular time is extremely small. In other words, the probability of internal blocking (for this level of traffic) in the switching network of Figure 3.13 is close to zero. The point to note is that the number of crosspoints has been reduced from 1 000 000 to 200 000 without seriously affecting the performance of the switching network.

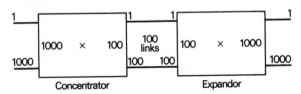

Figure 3.13 Use of two smaller switching matrices in place of one 1 000 × 1 000 matrix

The type of switching network shown in Figure 3.13 is said to concentrate the incoming traffic onto the 100 links and then expand it again on the outgoing side to give access to all outgoing lines. The first switching stage is therefore called a **concentrator,** and the second an **expandor.** To reduce the number of crosspoints still further, whilst retaining the ability to handle adequately 50 erlangs of traffic, the concentrator and expandor can be split into a number of smaller matrices, say 100 × 10 and 10 × 100, and each of these associated with a group of 100 lines. This technique is referred to as **grouping.** This is illustrated in Figure 3.14. The fact that each switching stage consists of 10 matrices is indicated by × 10 on the diagram; only the

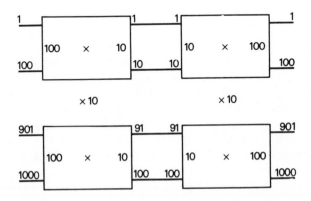

Figure 3.14 The use of smaller switching matrices by grouping lines into hundreds. With this arrangement it is not possible to interconnect different groups

91

first and last matrix in each stage are shown, and also only the first and last
link in each group of 10 links. Since the traffic per line is 0·05 erlang, each
group of 100 lines handles about 5 erlangs of traffic.

With the two switching stages interconnected as shown, it would be
impossible to connect a call between two lines in different groups. What has
to be done, therefore, is to provide a means of getting from one group of links
to another. This can be done using another switching stage, known as a
distributor, as shown in Figure 3.15. This is a 100 × 100 matrix which can
switch any incoming link to any outgoing one. The switching network shown
in Figure 3.15 contains 30 000 crosspoints and can perform exactly the same
function as the network with 200 000 crosspoints in Figure 3.13. However,
blocking can occur more easily in the network of Figure 3.15. If the first
switching stage is called A, the second B and the third C, we can refer to the
links as A-B links and B-C links, as appropriate. If there are 10 calls on an A

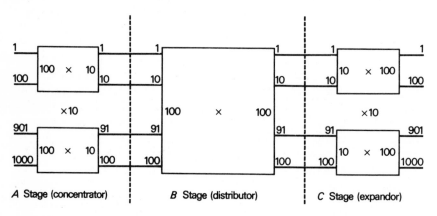

Figure 3.15 Use of distributor switching stage to make interconnection of
different groups possible

stage group, then all 10 A-B links for that group will be in use. Blocking will
then occur if an eleventh call occurs in that group. Similarly, if there are
more than 10 calls to be connected to a C stage group, then blocking will
occur. The probability of A-B link blocking is the probability of there being
10 simultaneous calls when 5 erlangs of traffic are offered. As you will see in
Chapter 4, this probability can be shown to be about 0·02. Similarly, the
probability of B-C link blocking is about 0·02. The probability of a given call
being successfully connected is thus the probability of no A-B link blocking
multiplied by the probability of no B-C link blocking, that is the product of
the two probabilities: 0·98 × 0·98 = 0·96. So, the probability of internal
blocking is (1—0·96) = 0·04. (This estimate assumes that A-B and B-C link
blocking can be considered independently. This is not strictly true in
practice. More precise calculations would show that the probability of

internal blocking is significantly higher than this, as described in Reference 34.) This probability of internal blocking may be acceptable for a given exchange. If not, then the number of links per group would have to be increased, and consequently also the dimensions of the matrices. For example, the three stages could use matrices of 100 × 15, 150 × 150, and 15 × 100.

To reduce the number of crosspoints still further, it is possible to extend the technique of grouping to the B stage. Figure 3.16 shows how the B stage can be split into 10 matrices of 10 × 10. Look carefully at the pattern of interconnections between the stages. Outlet 1 of group 1 at the A stage goes to inlet 1 of group 1 at the B stage. Outlet 2 of group 1 goes to inlet 1 of group 2. For group 1, therefore, outlet X goes to inlet 1 of group X. Furthermore, outlet 1 of group 2 goes to inlet 2 of group 1, and so on. Thus, in general, outlet X of group Y at the A stage, goes to inlet Y of group X at the B stage. This pattern of interconnection is called a **transposition.** In this case there is said to be a **full availability transposition,** that is every group in one stage has a link to every group in the previous stage. The transposition between the B stage and C stage is exactly similar.

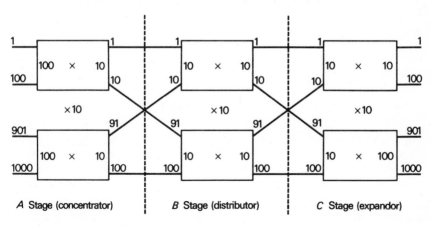

Figure 3.16 Splitting the distributor stage into ten groups

There are 10 possible paths from each inlet on an A stage matrix to any given outlet on a C stage matrix. Each one goes via a different B stage matrix. With 5 erlangs per A stage group, the traffic per A-B link is 0·5 erlang. This means that each link is busy half the time on average. B-C links are similarly busy half the time on average. Making similar assumptions about the independence of blocking in different stages as before, the probability of each of the 10 possible paths through the network being free, that is the A-B and B-C links being free in each case, is approximately 0·5 × 0·5, which is 0·25. The probability of each path not being available is thus 0·75. The probability of internal blocking is the probability that not one of

93

the 10 possible paths is available. This is $(0.75)^{10}$, which is about 0·06. So, this switching network, which contains only 21000 crosspoints, can handle 50 erlangs of traffic with a probability of blocking of 0·06. (This is a very crude calculation; a more rigorous analysis is given in Reference 34.)

As before, it would be possible to improve on this probability of blocking by increasing the number of links and the dimensions of the matrices. Another way of improving the probability of blocking, without substantially increasing the number of crosspoints, is to introduce another distributor stage, as shown in Figure 3.17. This gives almost as low a probability of blocking as the network with a 100 × 100 distributor matrix shown in Figure 3.15, yet has only 22000 crosspoints instead of 30000.

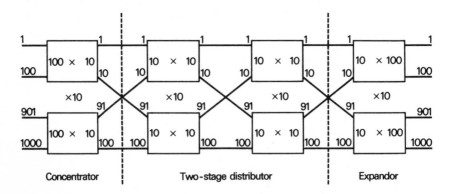

Figure 3.17 Use of a two-stage distributor

If the exchange had more than 1000 lines, then the use of a multi-stage distributor would be essential, in order to ensure a high probability of interconnecting a given inlet of the first switching stage with a given outlet of the last switching stage. For example, a practical switching network for a 10000 line exchange might use between six and eight switching stages in all. In general, the greater the number of lines, the more switching stages are needed for the optimum switching network (up to a certain number).

Until now no distinction has been made between local lines and trunks. However, there are two important differences. First, in connecting a call to a local line there is only one line that will do, namely the one which the user wants. However, when a call is being connected over a trunk route, any free trunk on that route is as good as any other. This means that the expandor is not essential for giving access to trunks, since the expandor is only really needed to make possible a choice of outgoing line.

The second point of difference is that traffic levels on trunks tend to be much higher than on local lines. The traffic per trunk may be 0·5 erlang or more, whereas the traffic per local line is usually 0·1 erlang or less. This means that the concentrator is not necessary for trunks; the traffic level on

94

trunks is already high enough for them to be connected direct to the first stage of the distributor. So, in a practical exchange design, trunks are normally connected as shown in Figure 3.18. Considered in isolation, the distributor switching stages would give a high probability of internal blocking for connecting a particular incoming trunk (or link from a concentrator matrix) to a particular outgoing trunk (or link to an expandor matrix). But the probability of internal blocking for connecting a particular trunk (or link) to any one of several trunks (or links) is much lower.

Note that, in Figure 3.18, the concentrator, distributor and expandor have been represented by three boxes; the individual switching matrices are not shown.

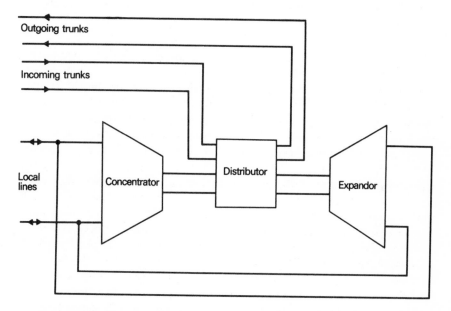

Figure 3.18 Practical arrangement of switching equipment. Trunks are connected direct to the distributor

The above description of some of the techniques involved in designing efficient switching networks has been necessarily brief. The subject is dealt with in greater depth in Reference 34.

3.4.3 Efficient control equipment structures

As mentioned earlier, to make efficient use of control equipment, it must be shared between several lines or calls in such a way that it is only brought into use on each call when it is needed. One method of achieving this is to use what is called **functional subdivision.** The control is broken down into a number of units, each one performing a few specific operations. The functions that a control has to perform may be summarised as follows:

(a) Recognition of the call-request signal.

(b) Reception and storage of the address information.

(c) Translation of the address to identify the required local line or trunk route and, where necessary, determine the routing for the call.

(d) Selection of a path through the switching equipment and operation of the crosspoints needed to establish that path.

(e) Sending signals (either local or inter-exchange) back down the incoming line and forward over the outgoing line.

(f) Monitoring the two lines to detect call-answered and clear signals.

(g) Clearing down the connection.

It is possible to break down the control into units which perform one or two of these seven basic control functions, or even just part of one of them. When this is done, three categories of control unit emerge. First, **per-line** units, that is units which must be permanently associated with one line and cannot be shared at all. In fact, only one unit is in this category. This is the unit which performs function (a) and which is called a **line unit.** It is permanently associated with one line because it has to be able to receive a call-request signal at any time. The second category is that of **per-call** units, that is units which must be associated with a particular call during the whole of the time it is in progress. A unit of this type is that used to perform function (f), and also, in some cases, function (g). This is called a **junctor** or **supervisory unit.** It is associated with a call throughout its duration because it has to be able to receive a clear signal at any time during a call. The third category is that of **per-set-up** units, that is units which are only needed while the call is being set up. Units to perform functions (b), (c), (d), and (e) are in this category. The names normally given to units performing each of these functions are: (b) **register,** (c) **translator,** (d) **marker,** and (e) **sender.** Where the sending of signals involves the reception of other signals, for example, a proceed-to-send signal from another exchange, the sender may be called a **sender/receiver.**

Figure 3.19 shows a typical arrangement of control units in a local exchange. The switching equipment is shown using the same representations of the concentrator, distributor and expandor that were used in Figure 3.18. We shall now go through the setting up of a call in this exchange. In doing so a number of options for the operation of the control units will emerge.

On detecting a call-request signal, a line unit sends a signal to the marker over an internal signalling path. The marker selects a free supervisory unit and connects the line to it by sending instructions over a control path to the switches in the concentrator. The marker then finds a free register and connects this to the supervisory unit by means of a special switching network, called an **access switching network.** This gives each call access to the several

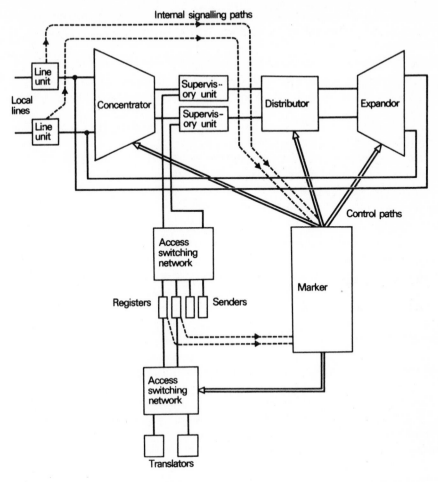

Figure 3.19 A typical arrangement of control units and switching equipment in a local exchange

registers and senders in the exchange and thus makes it possible to share these between a number of calls. The access switching network is similar in structure to a distributor. Once connected into the path, the register sends a proceed-to-send signal (for example dial tone) and then receives and stores the dialled digits.

Once a certain number of digits have been received, the register sends a signal to the marker over an internal signalling path. The marker selects a free translator and connects the register to the translator using a second access switching network. The register sends the stored digits to the translator. The translator translates these digits and the translated information is sent back to the register, where it is stored. The translator is then free to deal with another call. The time for which a translator is occupied on a particular call may be only a fraction of a second.

The register sends signals to the marker to instruct it to find a path through the distributor and expandor to the required line, and to operate the crosspoints necessary to set up this path. The register also ascertains whether or not the called line is free. There are three ways in which this can be done:

(1) A signal can be passed through the switching equipment from the line unit of the called line to the register.

(2) The marker can check the state of the line before setting up the path through the switching equipment.

(3) The translators can access the line units directly to determine whether a line is free or busy, and send this information to the registers.

If the line is busy, the marker may select a busy-signal sender, and connect this in place of the register. Alternatively, a busy-signal may be sent from the supervisory unit. If, however, the line is free, the register instructs the marker to connect the supervisory unit through the distributor and expandor to the called line. If necessary, a sender may be connected into the path by means of the access switching network.

If the call is to go via another exchange, the sender used is a digit signal sender which sends forward the address information to the next exchange. If the call is to a local line, an alert-terminal signal sender may be used. In some exchanges the sending of the alert-terminal signal is carried out by the supervisory unit.

Once the call is set up, the register is free to deal with another call. The total time for which the register is occupied is thus of the order of 10 seconds. The supervisory unit, however, remains associated with the call until the caller clears.

Cleardown of the connection is initiated by the supervisory unit. This may be done by a signal sent to the marker, or by the removal of an electrical condition on an internal signalling wire, connected through the switching network along with the speech path wires.

There are many other ways of structuring an exchange besides that just described. Control units which combine several of the control functions are sometimes used. Combined register-translators are used in some exchanges, and combined register-senders in others. However, there are two important points that apply to all exchange structures. First, the line unit is very simple; the supervisory unit is more complex; and the per-set-up control units, such as the register and the translator, are the most complex of all. Secondly, the number of control units needed is greater for the simpler units and smaller for the more complex units. For example, on a 1000 line exchange there might be 2 translators, 2 markers, 10 registers, 10 senders, 100 supervisory units, and 1000 line units.

An alternative method to functional subdivision for the structuring of the control equipment is the use of control units which are shared between calls in a different way. Instead of having, for example, a register which spends 10 seconds on one call, then 10 seconds on the next, a control unit containing fast electronic logic circuitry can spend a few milliseconds (or even micro-seconds) with one call, then another, and so on, dealing with a certain number of calls in rotation. It thus spends a number of short periods with each call at regular intervals, and carries out the control functions for each call in a piecemeal fashion as described below.

Many of the control units described earlier contained two distinct parts: signalling elements and control elements. The signalling elements are those parts of a control unit which send and receive local or inter-exchange signals. The control elements are those parts of the circuit which perform logical operations on the information received from the signalling elements and from other control units. The results of these logical operations are instructions to the signalling elements and information sent to other control units. The essential control functions can be described in terms of sequences of logical operations which can be carried out by a high-speed logic unit, shared between a number of signalling elements.

The structure of an exchange using time-shared electronic logic circuitry is similar to that of an exchange using functionally subdivided control, except that control units like the register are replaced by simple signalling units which send and receive local or inter-exchange signals, but perform no actual control function. These signalling units act as slaves to the time-shared logic. An example of this is as shown in Figure 3.20. Here it has been assumed that there is only one time-shared control for the whole exchange; this is often effectively the case in modern exchanges. Signals received by the various signalling units are fed into the control by means of a high-speed signal scanner, which scans the signalling units in rotation. Instructions, telling the signalling units to send signals, are delivered by means of a signal distributor. The switching equipment is controlled via a switching control interface, which gives the control access to each part of the switching equipment on a random, rather than sequential, basis.

In the last fifteen years this idea of a time-shared electronic control has developed into the concept of **stored program control** or **s.p.c.** for short. This involves the use of a digital computer as the time-shared control. Instead of its function being defined by its circuitry, the function of a computer is defined largely by the program which is stored in its memory. In an s.p.c. exchange the program in the computer causes it to behave exactly like a purpose-built control, but there are several advantages that stored program control has over other types of control.

First, it is possible to make changes in the way the exchange works by changing the program at any time. This greatly simplifies the introduction of new maintenance and monitoring procedures and new types of service for the user, during the lifetime of the exchange.

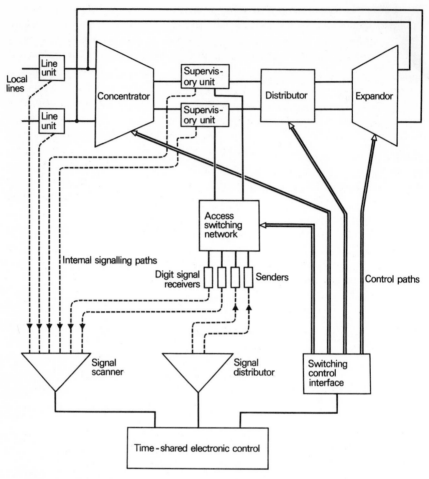

Figure 3.20 Possible structure of an exchange using time-shared electronic control

A second advantage of s.p.c. is that it opens up the possibility of much cheaper inter-exchange signalling. It is possible to implement common channel signalling for the trunks between two s.p.c. exchanges by providing a data link direct from the computer of one exchange to that of the other. No per-trunk signalling equipment is needed. The savings that result from this become greater the more trunks there are between the exchanges, since no extra signalling equipment is needed as the number of trunks increases.

A third advantage of s.p.c. is that the cost of performing the marker function in the computer is not greatly affected by the structure of the switching networks. This facilitates the use of many-stage switching networks which increases the crosspoint utilisation and traffic capacity of the exchange.

In other types of exchange it is normally necessary to compromise between the efficiency of the switching network and the cost of a marker; a marker which can select a path through an optimal network, containing many switching stages, would be extremely complex. With s.p.c. this complexity is transferred to the program and so its effect on the cost of the exchange is greatly reduced.

Although s.p.c. has significant advantages, it does introduce a number of problems. One is the cost of small exchanges. The basic cost of the computer, including its program store, is more or less independent of the size of the exchange. This means that small exchanges using s.p.c. are disproportionately expensive as compared with other types of exchange. Another problem which s.p.c. poses is that of reliability. If the computer breaks down, then the exchange stops functioning completely. An exchange whose control is divided into many units is inherently more secure because the failure of one unit only degrades the service given by the exchange very slightly. But a single component failure in the computer of an s.p.c. exchange is potentially able to put the exchange out of service. It is therefore necessary to make the computer able to operate in the presence of a certain number of faults.

The techniques used in fault-tolerant computers cannot be described here in any detail. The principle behind all fault-tolerant computers is that of **redundancy,** that is of providing more versions of each part of the computer than are needed for normal operation. Thus, several parts may become faulty and the computer continue to operate using the remaining fault-free parts. The simplest form which this kind of arrangement can take is that of a completely duplicated computer. The problem of achieving reliable control with s.p.c., and some of the possible solutions, are discussed in Reference 34.

As a final point about s.p.c., it should be noted that the cost of production of software (the computer programs) is a significant proportion of the cost of exchange development.

3.4.4 Types of switching equipment

The two most widely used types of switching equipment are called **strowger** (named after the Kansas City undertaker, Almon Strowger, who invented it in 1891) and **crossbar.** Before the advent of s.p.c., the structure of the control equipment tended to be closely tied to the type of switching equipment used in an exchange. There is thus a tendency for all designs of strowger exchange to be similar to one another, and all designs of crossbar to be similar to one another. However, the use of s.p.c. has helped to revolutionise the thinking of exchange designers. In non-s.p.c. exchanges there is rarely a clearly defined boundary between switching equipment and control equipment. But the computer in an s.p.c. exchange is so obviously separate from the switching equipment which it controls, that designers have become more able to see the switching equipment as a separate entity. Nowadays, one would therefore set out to design an exchange and choose, say, crossbar switching equipment because it was the most suitable for the job, rather than setting out to design a crossbar exchange as such.

A very brief description of the principles of operation of strowger, crossbar and other types of switching equipment will now be given. Strowger equipment uses crosspoints which consist of metallic contacts, brought into contact with one another by means of a rotary motion of one contact. The moving contact acts as one half of a number of crosspoints, the other half of each crosspoint being a fixed contact. Each switching device is called a **selector**, and has just one inlet. It thus represents a single row of a switching matrix. To create a complete matrix, a number of selectors have their outlets connected in parallel, as illustrated in Figure 3.21. This shows five selectors with five outlets each, connected in parallel to form a 5 × 5 switching matrix. Practical selectors have anything from 25 to 200 outlets, and any number of these may be connected in parallel to give the required number of inlets.

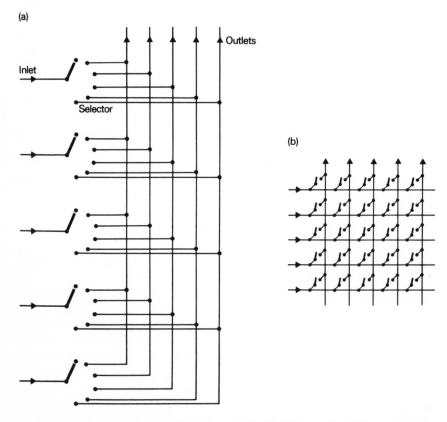

Figure 3.21 (a) The connection of several selectors in parallel to form a switching matrix. (b) Equivalent 5 × 5 switching matrix

There are two common types of strowger selector: a one-motion selector and a two-motion selector. The one-motion selector, or uniselector, is driven round by either a motor, or a solenoid and ratchet arrangement. The motor uniselector is stopped at the required outlet by means of a clutch mechanism. The solenoid uniselector, which is driven round by pulsed operation of the solenoid, is stopped by cessation of the pulsing current. A two-motion selector contains two solenoid mechanisms. One steps the moving contact vertically, and, when this has been done, the second solenoid steps it in a rotary manner in a horizontal plane. The fixed contacts are arranged in ten rows. The first stepping motion positions the moving contacts at the start of one of these rows, and the horizontal motion takes it round through the row to the required contact. A detailed description of the working of strowger selectors is given in Reference 29.

The important feature of the two-motion selector is that its vertical motion can be controlled by a digit signal from a dial, so that the row of contacts which is selected is determined by the digit. This makes the control of strowger equipment potentially very simple. Each selector normally contains a small control unit which is equivalent to part of a marker. When supplied with an instruction in the form of a digit, this control unit can select a free outlet in a particular row, the row being the one which is specified by the digit. A series of digits, sent one by one to each switching stage in turn, can thus select and set up a path through the switching equipment, the outgoing trunk or local line to which the connection is made being specified by those digits.

In the simplest strowger exchanges, the selectors are controlled directly by signals from the dial in the telephone, or by similar loop-disconnect digit signals sent over a trunk. This offers a very cheap design of exchange. Unfortunately, in multi-exchange areas, either the system has to be made somewhat inefficient, or else the user has to be burdened with a complex set of dialling codes. To make a strowger exchange more satisfactory from the user's point of view, a number of combined register-translators are used. These translate the digits dialled by the user into another set of digits, which control the operation of the selectors. An exchange of this type is known in Britain as a **director** exchange.

The simpler type of exchange, in which the strowger selectors are controlled directly from the dial, is referred to as a **non-director exchange.**

A crossbar switching mechanism, or crossbar switch as it is usually called, contains crosspoints which consist of pairs of metallic contacts, normally coated with silver. (Silver is the most suitable precious metal coating for contacts exposed to the air because both its oxide and sulphide are electrically conductive.) The contacts are arranged in a matrix, so that the switch itself resembles the diagramatic representation of a switching matrix. Along each row and column of the switch there is a metal bar, which can be rotated about $15°$ by the action of a solenoid. At the intersection of each row and column there is an intricate arrangement of levers, which causes the crosspoint contacts to be pushed together when the corresponding row and column bars are rotated. Each crosspoint can thus be selected by operating the appropriate row and column solenoids. In fact in most crossbar switches each horizontal bar serves two rows. The bar can be rotated one way to select one row, and the other way to select the other. A crossbar switch has typically a 10 × 20 crosspoint arrangement.

Once a crosspoint has been operated, the row bar can be released. In most crossbar switches the column solenoid has to remain operated to maintain the operation of the crosspoint, but more recently mechanically latching switches have been manufactured. These have the advantage of consuming no power once a path has been set up. Since a full description of a crossbar switch is rarely successful in conveying a clear picture of how it works to anyone who has not seen the real thing, it will not be attempted here. If you have a good visual imagination you might try Reference 29.

Crossbar switching equipment has for some time been used with control equipment whose structure is similar to that of the example exchange of Figure 3.19. More recently it has been used in s.p.c. exchanges. A number of other switches, based on the crossbar principle, have been produced recently, for example, the Miniswitch, which is small enough to fit on one printed circuit board (about 20 cm × 20 cm × 3 cm), and forms a 16 × 16 matrix.

Conventional relays have proved too expensive for use as crosspoints. However, **reed relays** have been found to be both cheaper and more convenient. A reed relay consists of a glass tube, surrounded by a coil, and containing two nickel-iron reeds, with gold plated tips, as shown in Figure 3.22. The reeds are springy and their tips are normally separated by a gap of about 0·13 mm. When a current is passed through the coil, the reeds become magnetised and, since the two tips become opposite magnetic poles, they are attracted together, thus creating an electrical contact. The glass tube is sealed and filled with nitrogen, to keep the contacts clean and to prevent oxidisation.

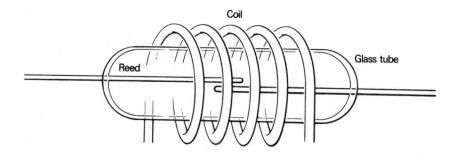

Figure 3.22 The basic features of a reed relay

The earlier types of reed relay needed a continuous current through the coil to keep them operated. Other types, called Ferreeds, have ferromagnetic plates inside the coil. A double coil is used. A current passed in the same direction through both halves causes magnetisation of the plates and operation of the relay. Currents passed in opposite directions through the two halves demagnetise the plates, and thus release the relay. A more recent type of reed relay, called the Remreed, uses ferromagnetic reeds, and thus behaves like the Ferreed, but is less bulky. Reed relays can be used to make up switching matrices of any required size.

Exchange designers have for many years been trying to produce an electronic crosspoint device suitable for use in exchanges, such as telephone exchanges, which carry analogue signals. It is extremely difficult, however, to make a device which has a high enough 'off' resistance. Each outlet from a switching network has thousands of paths to various inlets through switches which are in the off state and the accumulative signal leakage through these paths appears as a high noise level unless the off resistance of the crosspoints is at least 10^8 ohms. Also it is difficult to make electronic crosspoint devices that are sufficiently electrically linear to ensure that harmonic distortion of the signal is acceptably low. Furthermore, if they are to be used with a conventional telephone terminal, electronic crosspoints must be able to pass the d.c. needed for the microphone and the 75 volts a.c. ringing current of the bell. This is another requirement which is difficult to meet satisfactorily.

However, when the signals to be switched are binary digital, such as in a circuit switched data system, electronic devices make very satisfactory crosspoints. This is because linearity and off resistance do not matter where there are only two signal levels. The crosspoint devices in a digital exchange are therefore, in effect, simple logic AND gates, one of whose inputs forms the inlets of the crosspoint, and the other of which goes to the control. When the control input is set to a logical 1, the digital signals applied to the other input are reproduced at the gate output, and the crosspoint is thus on. With a logical 0 applied to the control input, the crosspoint is off. The use of electronic crosspoints in digital exchanges opens up the possibility of extremely efficient switching network designs, by using time-sharing techniques. These will now be described.

3.4.5 Time-shared digital crosspoints

Digital transmission links are normally connected to a digital exchange in multiplexed form, that is in groups of, typically, 30 active channels on a 2·048 Mbits s^{-1} highway. To connect a channel in one 30 channel transmission system to a channel in another system requires two operations. First, the channel must be shifted from the time-slot it occupies in one system, to the one occupied in the other system. Secondly, it must be switched through from one system to the other in the normal way. These two processes are referred to as **time switching** and **space switching.** For these to take place it is normally necessary to have all transmission systems that are connected to the exchange synchronised, so that all the time-slots correspond almost exactly.

The principle of these two switching processes is illustrated in Figure 3.23 which represents just four transmission systems with only four channels in each. The incoming channels of transmission systems 1 and 2 are switched to the various outgoing channels of systems 3 and 4. The diagram shows only one half of each connection; there is a corresponding other half for the other

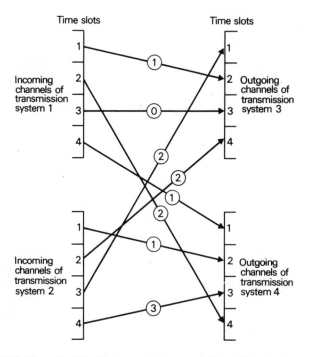

Figure 3.23 The principle of time and space switching. The numbers indicate the number of time-slots by which each channel must be delayed

direction of transmission, that is for the outgoing channels of systems 1 and 2, and incoming channels of systems 3 and 4. The time-switching process is represented by the circles, which contain the number of time-slots by which each channel has to be delayed. Note that, to get from one time-slot to an earlier one, the channel is delayed. For example, to get from time-slot 2 to time-slot 1, the channel is delayed by 3 time-slots. (Remember that the time-slots occur in rotation, so that slot 1 comes after slot 4.) Where the time-slots of the incoming and outgoing channels coincide, no time-slot shifting is needed.

A possible switching network is shown in Figure 3.24. The incoming highway for each transmission system is connected to an incoming time

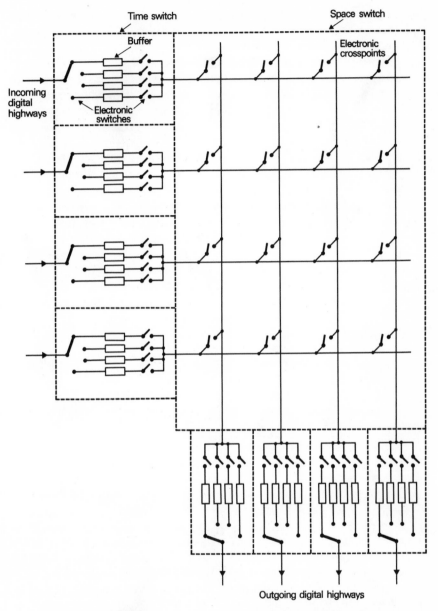

Figure 3.24 Example of a switching network for a digital exchange

switch. Only four time-slots have been shown for simplicity. An electronic switch in each incoming time switch places the bits from each time-slot (typically 8 bits) into a buffer. This is basically a shift-register. The buffers are filled in rotation, with one buffer allocated to each time-slot. A set of

107

electronic switches then removes the contents of each buffer during a later time-slot, and the bits pass down a highway into the space switch. At the same time an electronic crosspoint is operated in the space switch and this provides a path through to the required outgoing time switch. The bits thus pass into an outgoing time switch where an electronic switch is operated to direct the bits into the correct buffer for the required outgoing time-slot. An electronic switch, extracting the buffer contents in rotation, delivers the contents of each buffer to the outgoing highway of the transmission system in the appropriate time-slot.

The space-switching process is carried out by electronic crosspoints which are operated for just a few microseconds in the appropriate time-slots. They are thus shared between a number of concurrent calls. The outlets of several time switches can be multiplexed together to form highways carrying several hundred channels at a higher bit-rate. This means that the crosspoints are time-shared to an even greater extent. The fact that the space switch crosspoints are used so efficiently means that the space switching network can be designed for the convenience of the control; an efficient switching structure is not essential. This normally leads to the choice of a simple $N \times N$ matrix of crosspoints. In this example it has been assumed that the bits held in the buffer of an incoming time switch pass serially through the space switch. In practice a separate line may be used for each bit so that the space switch consists of, say, eight matrices acting in parallel, one for each bit.

The sort of arrangement described here is called a T-S-T network (short for time-space-time). Other structures, such as S-T-S, have been used, but T-S-T is the one used in most digital exchanges. A description of a practical digital exchange design can be found in Reference 34.

3.4.6 Message switched exchanges

To end this section, a brief description of some of the features of message switched exchanges will be given. This will be much less detailed than that of circuit switched exchanges for two reasons. First, the principles involved are much simpler but capable of being applied using a wide range of different practical arrangements. And secondly, the concepts involved are much less important to the design of telecommunication systems as a whole.

As mentioned in Chapter 2, the signalling processes in a message switched system are not as separate for the communication processes as in a circuit switched system. The required destination for each message is indicated by an address at the head of each message, this being in exactly the same form as the message itself. The function of the exchange can be described as follows:

(a) To examine the address at the head of the message.

(b) To determine whether the terminal is connected to the exchange in question or to a different one.

(c) To deliver the message to the appropriate transmission link as soon as that link is free.

The appropriate transmission link is either the local line of the required terminal (if it is connected to the exchange in question) or a trunk. The routing of messages is based on the same principles as the routing of calls in a circuit switched exchange using destination code routing. The message address includes the destination code, that is a code identifying the exchange to which the destination terminal is connected.

Messages are sent one after another down each transmission link and a queue of messages is formed for each link when messages cannot be sent immediately. The trunks in a message switched system may thus be almost continuously in use at certain times of the day. A typical exchange structure is shown in Figure 3.25. Incoming messages flow, bit by bit, into buffers associated with each transmission link. These hold a number of bits for a short period until the router has time to extract them. The router deals with

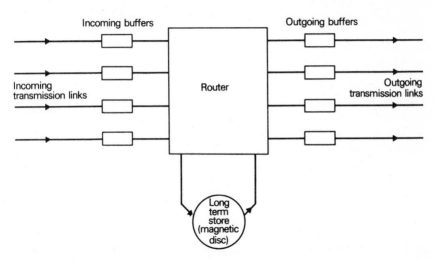

Figure 3.25 Basic features of a message switched exchange

each buffer in rotation. It extracts the bits from each buffer and transfers them to a long term store, such as a magnetic disc of the type used for computer memories. When a complete message has thus been assembled in part of this store, the router examines the address at the head of the message, and determines which outgoing transmission link the message is to go out on. The router then transfers the assembled message to a different part of the long term store, where messages queuing for the required transmission link are held.

As a second part of its operation, the router takes parts of messages from these queues, and places them in the appropriate outgoing buffers. Each time an outgoing buffer becomes empty, it is refilled with more of the same message, or the start of the next one.

The router must operate at very high speeds. It is effectively time-shared between three processes: collecting incoming messages, determining message routings and dispatching outgoing messages. A router would, in practice, consist of either a purpose-built piece of high-speed electronic circuitry or, more probably, a digital computer programmed to perform the required operations.

3.5 NEW IDEAS IN SYSTEM DESIGN

What has been described in this chapter consists mainly of well-established techniques and ideas. There are several important new ideas emerging at the present time. This chapter ends with a brief description of three of them.

The first and most important idea, is that of an **integrated digital system.** Digital switching equipment using time-shared electronic crosspoints tends to be much cheaper than analogue switching equipment. The overall cost of a digital system, plus the appropriate analogue-to-digital and digital-to-analogue convertors to convert speech into digital form using p.c.m., can be less than that of the equivalent analogue system. Because the digital transmission channel carries a p.c.m. representation of the analogue signal, there is no attenuation produced by transmission over any number of links and also the noise characteristics of digital transmission links and exchanges tend to be better than those of analogue links and exchanges. A telephone system in which at least the trunks and trunk exchanges are digital may therefore be cheaper and provide a more satisfactory service than an analogue system of the kind widely used at present. There is the additional advantage that the system can be shared between telephones and data terminals, thus producing an integrated telecommunication system serving many needs. The sort of data terminals that such a system could serve include teletypewriters, computers and digital facsimile machines. The most economic arrangement for the positioning of the analogue/digital convertors is still uncertain. Present trends in the relative costs of analogue and digital transmission and switching equipment seem to indicate that it may eventually be economic to put the convertors in the terminal. This would make it possible to have local lines carrying, say, 80 kbits s^{-1} digital signals and completely compatible between speech and data.

The second new idea is that of **remote control**. In modern exchanges, particularly s.p.c. ones, there is a tendency for the control equipment to become easily separable from the switching equipment. By taking all the signals that pass between the control equipment and switching equipment and multiplexing them, in binary digital form, onto one digital channel, it is possible to have the control equipment a considerable distance from the rest of the exchange. This arrangement is called remote control and the digital channel between the control equipment and switching equipment is known as a control signalling link. Remote control may be used to keep down the initial cost of s.p.c. exchanges. What can be done is to install only the switching parts of a new s.p.c. exchange and to connect these by a control signalling link to the computer of an already established s.p.c. exchange. The computer is thus shared between its own exchange and the remote exchange. When the remotely controlled exchange has grown to a point where the shared computer is becoming overworked, the remote exchange is given its own computer.

The third idea is that of **area control.** Area control is an extension of the use of remote control. What it involves is the use of a powerful and highly reliable computer installation to control all the local exchanges and the primary trunk exchange in one area via control signalling links. Possible reasons for using area control are as follows. First, all the information which is used in translation processes in all the exchanges may be stored in the central computer, and may therefore become easier to update when required. Secondly, the central computer has a complete overview of the area and may thus be able to analyse traffic patterns and diagnose faults more easily than a number of computers dealing with one exchange each. Also, details of traffic and faults, together with information for the preparation of users' bills, can be recorded at the central computer, for example, on magnetic tape or punched cards. Thirdly, special services, which would involve the use of a complex computer program, but which would be used by only a few users, can be provided more efficiently by the use of a single copy of the program in the area computer. (This is not necessarily the case with regularly used pieces of program.)

With the advent of the minicomputer it has become economically feasible to adopt a semi-remote control approach to area control. A minicomputer can be used to perform some, but not all, of the control functions in a local exchange. The other functions, such as translation, billing, fault recording and traffic recording, can be carried out by a central computer over a control signalling link between the local computer and the central computer. This results in only functions which would be expensive and inconvenient to perform locally being performed centrally, while functions which are performed most efficiently locally are dealt with by the local computer. Experiments are being carried out to find the optimum distribution of control functions between the local and central computers.

Chapter 4

Switched telecommunication systems: design and planning

INTRODUCTION

Decisions about which techniques and technologies should be used in a particular element of a switched system and decisions about the detailed structure of the system are not generally clear-cut issues. Many of the decisions depend on a comparison of alternatives involving a number of factors which cannot easily be reduced to a common parameter such as cost. Much of Chapters 2 and 3 has therefore been of a descriptive nature, containing an account of techniques that have proved useful in practice, rather than an account of the reasons underlying the choice of particular techniques. In this chapter an attempt will be made to introduce you to some of the factors which may be taken into account in the design and planning of switched systems. Some of these factors are basically qualitative and some are quantitative. A choice or decision usually involves weighing up a mixture of qualitative and quantitative factors. Only where all the relevant factors are reducible to quantitative terms can an optimal solution be found. In most cases, however, there will be no optimal solution; no solution will have a combination of advantages and drawbacks which is demonstrably more desirable than any other. In such a situation an element of human judgement is necessary.

In this chapter we shall look first at some general aspects of the design of telecommunication equipment for large switched systems. Then we shall look at various topics in the planning of switched systems which are related, in a fairly direct way, to the service given to the user. The aspects of service given to the user which you should bear in mind when reading this chapter are:

(a) Availability of service, that is whether users can be given a terminal within a reasonable time of asking for one.

(b) Quality of service, as indicated first by difficulties or delays involved in using the system, such as the blocking of calls, and secondly, by the acceptability of the transmission channel in terms of attenuation, noise, and so on.

(c) Reliability of service, that is the extent to which users can depend upon their terminals and the system as a whole to be working at any time.

(d) Cost of service. (This is related somewhat indirectly to the effectiveness of the system planning, though the efficiency with which resources are used in providing the service must be reflected in the cost of the service over a long period of time.)

The aspects of design and planning discussed in this chapter are only some of those that may have to be taken into account in a particular situation.

The chapter ends with an overview of the British telephone system. This is included to give some idea of the size and complexity of a switched system of this kind, and the variety of techniques and technologies that are likely to be found in a system that has evolved over many years.

4.1 DESIGN OF TELECOMMUNICATION EQUIPMENT

In a large switched telecommunication system, such as a public telephone system, the design of equipment is dominated by two factors: **compatibility** and **modularity.**

Because such a system contains a large number of different elements interacting with one another in a complex manner, it is almost impossible to take a new element, which has been designed without any regard to the existing system, and to graft it satisfactorily onto the system. The way in which the new element will interact with the rest of the system must be thoroughly examined at every stage of design. This is particularly important when a new version of an existing element is to be introduced, for example, a new version of a particular element using more modern components. The new element must have an interface with the rest of the system which corresponds with the interface of the old element, and it must be able to perform at least the same functions as the old element. In other words, it is important to ensure that there is both interface compatibility and functional compatibility.

The result of the need for compatibility is that new versions of elements of the system are generally more costly than their counterparts would be in a completely new system, and they may even be more costly than the old element using older technology. The need for compatibility with the existing system therefore produces a barrier against technical innovation. A new technology is usually only adopted in the following cases:

(a) When the new components are so much cheaper that, in spite of the extra costs of achieving compatibility, the new element is cheaper than the old one.

(b) When the newer technology has substantial operational advantages which outweigh the initial cost considerations.

(c) When replacement components for repairing the elements based on the older technology are becoming difficult or impossible to obtain.

The barrier against the use of new technologies would be removed if the existing system were scrapped and completely replaced by a new one using the new technologies throughout: this is not normally feasible for the size of system we are considering. Not only would this involve an enormous waste of resources from scrapping equipment which still has a number of years of useful life left, but also it would be physically impossible to manufacture and install enough equipment before that, too, was out of date. Large telecommunication systems are thus said to exhibit a natural **inertia** with regard to technological change.

The second important factor in the design of telecommunication equipment for a switched system is modularity. In order to ensure that it will be possible to keep the amount of equipment close to the amount which will be needed, it must be possible to add to the equipment already in the system in reasonably small lumps, or modules. In other words, the equipment must be constructed so that a number of modules can be brought together to create an installation of any required size and then, as the system needs to grow, more modules can be added to increase the capacity of the equipment. If the equipment is not sufficiently modular, there will be a large proportion of equipment lying idle from the time a module is added, to the time when demand begins to catch up with the added capacity of the new module. This is wasteful of resources and would tend to make the equipment less economically attractive.

4.2 DEMAND FOR SERVICE

The demand for a telecommunication service has two aspects: the number of terminals needed, and the amount of traffic to be handled. These two are, of course, interrelated. Each user generates a certain amount of traffic. The use which each user makes of the system is often described by means of a **calling rate** for his terminal. This is the average number of calls he makes per hour, usually measured during the busy hour. Thus, a calling rate of 2 calls per hour, with a mean holding time of 3 minutes for calls, represents an average of 6 minutes use of the system per hour, that is, $0 \cdot 1$ erlang of originating traffic. There are usually also incoming calls to each terminal. These are described in terms of an incoming call rate, and a corresponding level of terminating traffic. For most terminals the originating and terminating traffic levels are roughly equal. Typical values for the originating traffic per terminal are, in a telephone system, $0 \cdot 1$ erlang for business telephones and $0 \cdot 05$ erlang for residential telephones.

The forecasting of the demand for a service, in terms of the number of new terminals that will be needed, is normally done on two levels. First, the trends for the whole system are examined, so that the total number of new terminals can be predicted. This prediction can take account of factors such as the population growth, the predicted state of the economy, the cost of travel which the use of the service might replace, and so on. Secondly, the number of new terminals that are likely to be asked for is forecasted on a local basis. Proposed building developments such as offices, factories and housing estates are taken into account. It is often useful to define an average **penetration** of the service for each type of building. For example, the penetration for houses can be expressed as the average number of telephones per household. (In Britain penetrations of much less than 1·0 are usual, but in North America penetrations greater than 1·0 are found in some residential areas; some households have second telephones for teenage children.) Local predictions are then gathered together and adjusted so that the total predicted number of terminals corresponds with the prediction for the system as a whole. The forecasts of the number of new terminals that will be needed are then used in two ways. They are used directly to determine the quantities of local exchange equipment and the number of local lines that will be needed, and they are used in predicting the growth in traffic carried by the system.

The forecasting of traffic growth, besides being based on the predicted growth in the number of terminals, is also based on trends in calling rates and on traffic growth patterns for previous years in various parts of the system. The measurement of traffic in various parts of the system is thus necessary to provide information on which system planning can be based. We shall look briefly at one of the possible methods that can be used to measure traffic. Figure 4.1 shows a traffic measuring device, or **traffic recorder** as it is more often called, which can be used to measure traffic in parts of a circuit

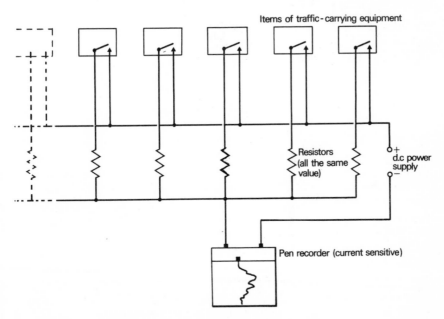

Figure 4.1 Example of a traffic recorder

switched system. It is assumed that each item of equipment that can carry one call has some sort of switch associated with it, this switch being on when a call is in progress, and off when there is no call.

Each switch, in the items of equipment to be monitored, is connected via a resistor to a d.c. power supply and a current-sensitive pen recorder. The deflection of the pen on the moving paper is proportional to the current flowing, and hence to the number of items of equipment in which the switch is on. The deflection of the pen is thus proportional to the instantaneous traffic level, that is to the number of calls in progress. If such a device is left connected to, for example, a number of traffic carrying devices in an exchange for 24 hours, then the result might be like the chart shown in Figure 4.2. This pattern would be typical of a weekday for a local exchange with both business and residential telephones connected to it. You can see

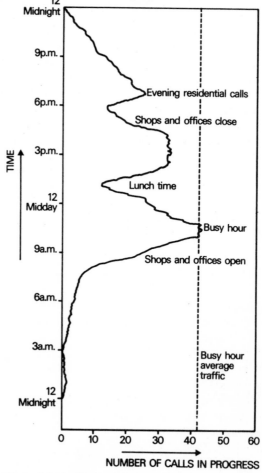

Figure 4.2 Traffic recorder chart for 24-hour period

that the busy hour is around 10.30 a.m. The busy hour average traffic has been indicated by a dotted line.

Using a traffic recording technique such as this, or some alternative method, it is possible to identify the busy hour for a particular part of the system, and measure the busy hour average traffic. Records of this sort can be made for exchanges as a whole, parts of exchanges and trunk routes. By plotting graphs of the busy hour average traffic, as measured every few months, the rate of growth (or decline) in traffic can be determined and, taking into account this and other factors, the traffic level at some time in the future can thus be predicted.

For a given traffic level, it is possible to calculate the number of items of equipment, such as trunks or parts of an exchange, that will be needed to achieve a given probability of call blocking in the part of the circuit switched system in question. The theory on which this calculation is based was developed by Agner Erlang (see Chapter 2, Section 2.9). The main points of this theory are dealt with later.

An aim of systems planning is to ensure that there is neither too much nor too little equipment for the expected demand. If there is too much, then resources are wasted. If there is too little then demand will not be met. The result of there being insufficient equipment for the provision of terminals where they are needed is usually a waiting list for terminals. The result of there being insufficient traffic capacity depends on the type of system. In a circuit switched system, in which calls that cannot be connected are blocked, an unacceptably large proportion of calls will be blocked. In systems, such as message switched systems, which involve queuing, the result is that the queuing delays become unacceptably great. In order to describe aspects of the quality of service concerned with the difficulties or delays involved in using the system, the system performance may be expressed in terms of one or several **grades of service.**

4.3 GRADES OF SERVICE

A grade of service usually expresses the probability of something not happening. For example, the following are commonly used grades of service in a circuit switched system:

(a) The probability of call blocking, that is the probability of a call not being connected at all.

(b) The probability of excessive delay in receiving dial tone, that is the probability of the user not receiving dial tone within a certain time after lifting the telephone handset.

(c) The probability of excessive delay in call set-up, that is the probability of the user not getting a tone within a certain time after completion of dialling.

Neglecting the effects of faults, each of these is related to the quantities of equipment provided and the traffic that has to be handled. To illustrate some of the principles involved in matching the equipment quantities to the traffic, the probability of call blocking will be used as an example.

As mentioned earlier, one aim of system planning is to make sure that there is neither too much nor too little equipment for the expected traffic. If there is too much, then resources are wasted. If there is too little, then one result may be that an excessive number of calls are blocked. It is important to achieve, as nearly as possible, a target probability of call blocking for each part of the system. This target is normally set by the administration and represents what is economically achievable. The target probabilities of call blocking might be expressed in terms of a target for each type of call, as shown in Table 4.1.

Table 4.1 *Examples of target probabilities of call blocking for various types of call*

Type of call	Target probabilities of call blocking
Between terminals on the same local exchange	0·02
Between terminals on different local exchanges, via one direct trunk	0·04
Between different local exchanges via one trunk exchange	0·05
Between different local exchanges via two trunk exchanges	0·07

These targets can be broken down into target probabilities of call blocking for parts of the system by using the fact that, if the probabilities of blocking in the various parts of the system are B_1, B_2, B_3, and so on, then the probability of the call not being blocked anywhere in the system is:

$$(1 - B_1)(1 - B_2)(1 - B_3) \ldots \tag{4.1}$$

In other words, the overall probability of blocking is:

$$1 - [(1 - B_1)(1 - B_2)(1 - B_3) \ldots] \tag{4.2}$$

If all the values of B are small (less than about 0·05), then this approximates to:

$$B_1 + B_2 + B_3 + \ldots \tag{4.3}$$

So, the overall probabilities of call blocking can be broken down into components for individual exchanges and trunk routes as shown in Table 4.2.

These are then the target probabilities of call blocking that are used in determining the equipment quantities in each part of the system, for

Table 4.2 *The result of breaking down the probabilities of call blocking into components for each part of the system*

Part of system	Target probabilities of call blocking
Local exchange, terminal-to-terminal connection	0·02
Local exchange, terminal-to-trunk and trunk-to-terminal connection	0·01
Trunk route between two local exchanges	0·02
Trunk route from local exchange to primary trunk exchange	0·01
Trunk exchange, trunk-to-trunk connection	0·01
Trunk route between trunk exchanges	0·01

example, the number of trunks on a particular trunk route. We shall now look at how the probability of call blocking can be related mathematically to the equipment quantities and the traffic to be handled.

4.3.1 Basic traffic theory

In Chapter 2 it was shown that the rate at which new calls arrive, a is related to the traffic, E, by the expression:

$$E = ah \text{ erlangs} \tag{4.4}$$

where h is the mean holding time of calls. The value of a was taken to be the rate at which new calls are connected. Because of blocking, in practice not all attempted calls are connected. The proportion connected depends on the probability of call blocking and on the action taken by users when they receive a signal indicating that their attempted call has been blocked. For example, users may try another attempt immediately, or try again a few minutes later, or give up completely. For simplicity, it will be assumed that users give up after one blocked call attempt. If this is so, Equation (4.4) can be used to relate the rate of arrival of all calls, including those that are blocked, to a value of E which we shall call the **offered traffic.** This is the traffic that there would be if all calls were connected. Because some calls are blocked, the **carried traffic** is less than this.

The probability per unit time of a call arriving, for a short interval of time, is equal to the rate of arrival of calls, a. The probability per unit time of a call ending depends, at any particular time, on the number of calls in progress, which we shall refer to as k. Since the average duration of each call is h, for a small interval of time the probability per unit time of each call ending is $1/h$. Therefore, when there are k calls in progress, the probability per unit time of a call ending is $k \times (1/h) = k/h$. If there are N items of equipment each of which can carry one call, we can describe each of the possible situations, such as no calls in progress, one call in progress, and so on, as a state. We can then say that the probability of a state existing, with k calls in progress, is

119

$P(k)$. Since there are only N items of equipment there cannot be more than N calls, so $P(k) = 0$ for $k > N$. Also, there cannot be a negative number of calls, so $P(k) = 0$ for $k < 0$. The sum of the probabilities of all possible states must be equal to 1, so that:

$$\sum_{k=0}^{k=N} P(k) = 1 \qquad (4.5)$$

In practice, the instantaneous traffic level, and hence also the call arrival rate, fluctuate, so the instantaneous values of $P(k)$ are not constant. However, for the purposes of this calculation, the values of E and a will be taken as constant and equal to the busy hour average values. The values of $P(k)$ will then be constant. This situation is described as one of **statistical equilibrium.** If each value of $P(k)$ is constant, the probability per unit time of entering each state must be equal to the probability per unit time of leaving it.

It is highly unlikely that two calls will start or finish at exactly the same instant, so it can be assumed that state k is entered only from state $(k-1)$, by a call arriving, or from state $(k+1)$, by a call ending. The probability of being in state $(k-1)$ is $P(k-1)$, and the probability per unit time of a call arriving is a, so the probability per unit time of entering state k from state $(k-1)$ is $P(k-1) a$. The probability per unit time of entering state k from state $(k+1)$ is $P(k+1) (k+1)/h$ since the probability of being in state $(k+1)$ is $P(k+1)$, and the probability of a call ending in that state is $(k+1)/h$. The total probability per unit time of entering state k is thus given by:

$$P(k-1) a + P(k+1) \frac{k+1}{h} \qquad (4.6)$$

By a similar argument, the probability per unit time of leaving state k is:

$$P(k) a + P(k) \frac{k}{h} \qquad (4.7)$$

The condition for statistical equilibrium is thus:

$$P(k-1) a + P(k+1) \frac{k+1}{h} = P(k) a + P(k) \frac{k}{h} \qquad (4.8)$$

Since $P(k) = 0$ for $k < 0$, for $k = 0$ Equation (4.8) becomes:

$$P(1) \frac{1}{h} = P(0) a \qquad (4.9)$$

For $k = 1$, Equation (4.8) becomes:

$$P(0)a + P(2)\frac{2}{h} = P(1)a + P(1)\frac{1}{h} \tag{4.10}$$

Substituting in this for $P(1)$ from Equation (4.9) gives $P(2)$ in terms of $P(0)$:

$$P(2) = P(0)\frac{a^2\,h^2}{2} \tag{4.11}$$

Putting $k = 2$ in Equation (4.8) and then substituting for $P(2)$ from Equation (4.11) gives:

$$P(3) = P(0)\frac{a^3\,h^3}{2\times3} \tag{4.12}$$

Carrying on like this up to $k = N$ shows that, in general:

$$P(k) = P(0)\frac{a^k\,h^k}{k!} \tag{4.13}$$

From equation (4.5):

$$P(0) = \frac{1}{\displaystyle\sum_{k=0}^{k=N}\frac{a^k\,h^k}{k!}} \tag{4.14}$$

so that:

$$P(k) = \frac{\dfrac{a^k\,h^k}{k!}}{\displaystyle\sum_{k=0}^{k=N}\frac{a^k\,h^k}{k!}} \tag{4.15}$$

Using equation (4.4), we can now replace (ah) by E, giving:

$$P(k) = \frac{\dfrac{E^k}{k!}}{\displaystyle\sum_{k=0}^{k=N}\frac{E^k}{k!}} \tag{4.16}$$

121

This is known as the Erlang probability distribution. It gives the probability of there being k calls in progress at any given time. What we want to know is the relationship between E, N and the probability of call blocking, B. B is the probability that all N items of equipment are in use, which is $P(N)$. Therefore:

$$B = P(N) = \frac{\dfrac{E^N}{N!}}{\displaystyle\sum_{k=0}^{k=N} \dfrac{E^k}{k!}} \tag{4.17}$$

This is one of the expressions on which the planning of parts of a switched system which carry traffic is based. For a given target probability of call blocking it can be used to find the minimum number of equipment items needed for the expected traffic. For example, Table 4.3 shows the number of items needed to achieve a probability of call blocking of 0·01 with various offered traffic levels.

Table 4.3 *The number of equipment items needed to achieve $B \leqslant 0\cdot01$ for various offered traffic levels*

Offered traffic (erlangs)	Items needed to make $B \leqslant 0\cdot01$
0·1	2
0·5	4
1·0	5
5·0	11
10·0	18
50·0	63
100·0	117

Note that the average traffic carried per item of equipment is greater the more traffic is carried. In other words, larger traffic levels can be handled more efficiently than lower ones. However, as you will see later, other factors have to be considered which limit the efficiency of handling traffic levels in excess of about 50 erlangs.

4.3.2 Traffic overload

The assumption made in the above calculation, that users whose call attempts are blocked do not try another attempt, leads to valid results when the proportion of blocked calls is small. Calculations based on other assumptions, for example, that all users keep re-dialling until their calls are successfully connected, give results that are only slightly different for low probabilities of call blocking. However, if the offered traffic becomes significantly greater than that for which the number of items of equipment was chosen, not only does B increase markedly, but also the differences in the value of B as predicted by theories with different assumptions about user behaviour become significant. No simple mathematical description of user behaviour has been found to be completely satisfactory; some of the methods of calculating B using various assumptions are described in Reference 34. The point to note is that, when there are a significant number of repeat-attempt calls, a is no longer equal to (E/h) because, as well as the new calls arriving at the equipment, there will also be a number of repeat-attempt calls. The true value of a is therefore greater than indicated by the offered traffic, so B becomes greater. In effect, the new calls have to compete with repeat-attempt calls for use of the equipment, so the probability of any given attempt being blocked increases.

If, for instance, a trunk route of 18 trunks, normally handling 10 erlangs, which gives $B = 0\cdot007$, is offered 11 erlangs of traffic, Equation (4.17) predicts that B will rise to $0\cdot015$. In practice, because of repeat attempts, the actual value of B might be about $0\cdot016$. Thus, an increase in the offered traffic of 10 per cent in this case makes B worse by a factor of about 2. With 20 per cent more traffic than normal, that is with 12 erlangs offered, B becomes about $0\cdot03$, 4 times the normal value. In other words, if the traffic in a particular part of the system becomes greater than that for which it has been designed, B rapidly becomes worse than the target value. As mentioned in Chapter 2, this situation is described as congestion, and it arises because of **traffic overload.** This can come about gradually, if the forecasts of traffic on which the planning has been based were wrong, or suddenly, either because of some event which causes a sudden increase in traffic, such as a disaster of some sort, or a fault in the system which causes traffic to be transferred from one part of the system to another.

If we consider the case of a part of the system where there are a greater number of items of equipment, such as a large trunk route, the effects of traffic overload turn out to be more marked. For example, a trunk route of 117 trunks, normally handling 100 erlangs, giving $B = 0\cdot01$, gives $B = 0\cdot05$ with 10 per cent overload, and $B = 0\cdot13$ with 20 per cent overload. That is, with 20 per cent overload, B worsens by a factor of 13, as compared with a factor of 4 in the case of a route of 18 trunks. The tendency of trunk routes and other parts of the system to be more susceptible to traffic overload the more traffic they carry normally, means that it may be necessary to provide more equipment than is needed for the target value of B in such cases. The value of B under normal conditions is thus better than the target value.

From Table 4.3 you can see that the efficiency of a trunk route increases the more traffic it carries. To carry 1 erlang, with a target value for B of $0\cdot01$, needs 5 trunks, so the traffic per trunk is $0\cdot20$ erlang. With 10 erlangs, the traffic per trunk is $0\cdot56$. However, as one considers larger routes, the overload performance of the route becomes important and it becomes necessary to have more trunks than are needed to achieve the target value of

B under normal conditions. The result of this is that it is not normally possible to achieve an average traffic level per trunk in excess of about $0\cdot8$ erlang, so the efficiency of a route does not significantly increase when more traffic is carried above about 50 erlangs.

4.3.3 Alternative routing

The way that exchanges are arranged into a hierarchy was described in Chapter 2. There can be three reasons for having a hierarchy of this kind:

(a) To ensure that a routing exists for every call.
(b) To limit the maximum number of transmission links involved on calls and thus keep transmission impairments within certain limits.
(c) To facilitate the use of alternative routing.

We shall look at (b) later. In this section we shall examine some aspects of alternative routing that have to be taken into account in planning a switched system.

The use of alternative routing, that is the use of a routing other than the basic routing when a route in the basic routing is completely busy, can, under a limited range of conditions, improve the overall efficiency of the system. It must be emphasised that the use of alternative routing outside this range of conditions decreases the efficiency of the system. We shall first look at the conditions required for alternative routing to increase the efficiency of the system.

In general, the use of an alternative routing implies the use of a greater number of transmission links and exchanges to form the transmission channel. The cost of connecting a call is thus greater for the alternative routing. This is because, first, the cost of a greater number of transmission links is more, even if their total length is the same, since the cost of the equipment at the end of each extra link in the alternative routing is added to the cost of the links themselves. Secondly, the alternative routing is usually less direct, so this adds further to its cost. And thirdly, extra switching costs are involved in the alternative routing as it passes through more exchanges. Nevertheless, the overall cost of connecting calls, taking into account the costs of calls connected both using the basic routing and using the alternative routing, may be less than the cost when only the basic routing is used. This can happen because, as a result of splitting the traffic between the basic routing and the alternative routing, all the routes involved can be used more efficiently. In other words, the average traffic per trunk can be made higher on all routes. An increase in the traffic per trunk means that the cost per erlang is less. The way in which this increased efficiency comes about is as follows.

Considering the system as a whole, the effect of alternative routing is to offer high traffic levels to certain auxiliary routes, which are referred to as

high-usage routes, and to allow traffic which cannot be taken by high-usage routes to spill over onto other routes. Routes which are capable of handling all the offered traffic without relying on an alternative route are called **fully-provided routes.** High usage routes generally consist of only a few trunks and are efficient by definition; the average traffic per trunk is considerably higher than if they had to handle all the offered traffic. The fully-provided routes which handle the spill-over traffic of many high-usage routes consist of many tens of trunks; they are efficient because of the large number of trunks, carrying an average of about 0·8 erlang per trunk (as dictated by the overload performance requirements).

In planning the trunk routes of a switched system, it is possible to predict the potential savings, if any, that can be obtained from using alternative routing in a given part of the system, and to roughly maximise these savings. The main consideration in doing this is the number of trunks on the high-usage routes. For example, suppose that the traffic offered to a particular high-usage route is E_0 erlangs, and that, using alternative routing, E_1 erlangs are carried by the high-usage route, the alternative routing taking $E_2 = E_0 - E_1$ erlangs of spill-over traffic. By comparing the costs of the routes involved in the alternative routing, and taking account of the cost of the extra switching involved on the alternative routing, it is possible to derive an approximate cost ratio, A, indicating the ratio of the cost of a connection over the alternative routing to the cost of a connection over the high-usage route. The optimum number of trunks on the high-usage route, which we shall call H, is the number for which the cost of providing the H trunks plus the cost of the necessary capacity on the alternative routing is a minimum. Taking the cost of one trunk on the high-usage route as a unit of cost, the cost of H trunks is H, and the cost of the extra capacity on the alternative routing is roughly $AE_2/0·8$. This is because the alternative routing, being made up of routes with many trunks each, carries roughly 0·8 erlangs per trunk, so to carry E_2 erlangs requires $E_2/0·8$ trunks. Therefore, the optimum value for H is one for which ($H + AE_2/0·8$) is minimised. Note that E_2 is a function of H; the greater is H, the more traffic is carried on the high-usage route, and hence the smaller is E_2.

There are thus two parameters, E_0 and A, which determine what is the most efficient arrangement. For particular values of these, one of three conditions may apply:

(a) Minimising the overall costs gives $H = 0$. In this case there should be no route provided between the exchanges concerned. All calls should be routed over the routing that was considered as the alternative.

(b) Minimising the overall costs gives a value of H for which E_2 is not negligible. In this case H trunks should be provided on the high-usage route and the spill-over traffic, E_2 should be carried by the alternative routing.

(c) Minimising the overall costs implies that E_2 tends to zero. In this case, alternative routing should not be used, and the route should be a fully-provided one capable of handling E_0 erlangs.

A simple way of deriving the optimum value for H is as follows. It is assumed that calls are offered to trunks on the high-usage route in turn, so that the second trunk is only used if the first is busy, and the third only if the second is busy, and so on. (Some types of exchange actually do this.) The traffic carried by the first trunk can then be calculated by finding the average proportion of time it is busy. This is done by putting $N=1$ and $E=E_0$ in Equation (4.17). The result, which we shall call e_1, is the traffic carried by the first trunk. By putting $N=1$ and $E=E_0-e_1$, the traffic carried by the second trunk, e_2, is calculated. Putting $E=E_0-e_1-e_2$, e_3 is calculated, and so on. In terms of cost per erlang, the use of the alternative routing, where about 0.8 erlang per trunk is carried, is equivalent to the use of a trunk on the high-usage route carrying $0.8/A$ erlang, since trunks on the high-usage route are cheaper by a factor A. The optimum value for H is therefore that for which $e_H > 0.8/A > e_{H+1}$. The value of H thus determined can be shown to be optimal no matter what arrangements are used for offering calls to the high-usage route, as long as calls are only routed over the alternative routing when all the high-usage trunks are busy.

To achieve efficient operation of a system requires more than just the optimisation of the number of trunks on each route. It is also necessary to ensure that the way in which exchanges carry out alternative routing is very carefully regulated. The danger is that exchanges acting completely independently in choosing routes from a number of alternatives may cause the call to be routed up and down the hierarchy many times, thus using a larger number of trunks than the backbone routing. Worse still, a call may be routed round in circles so that it uses up tens or even hundreds of trunks. To prevent this happening, each exchange must be made to act according to a set of rules. These rules dictate what routes can be used and in what order they should be tried, given a call's destination and the route on which it arrived at the exchange. Routing rules of this kind are discussed in detail in Reference 34.

4.4 TRANSMISSION PLANS

All aspects of the quality of service represent a compromise between the quality and cost of the service. Just as this leads to the choice of target probabilities of call blocking, so it also leads to a transmission plan. On a transmission channel involving a number of transmission links connected end to end, the accumulative impairments of the transmitted information by the several links will be considered, by some users, to be such as to make the transmission channel as a whole unsatisfactory. For any given transmission channel, there are three points to consider:

(a) The nominal value of the impairments caused by the transmission links.

(b) The deviation of the actual values of the impairments from their nominal values.

(c) The opinion of the users about the quality of the transmission channel.

The nominal values of the impairments are those ascribed to the transmission links by the transmission plan.

In a particular system there will be a number of types of impairment which the planner has to take account of. For example, in a telephone system the size of the British one, the main impairment which calls for careful planning in order to limit its effects is attenuation. Noise is also, to a certain extent, a problem, particularly impulsive noise caused by switching equipment in exchanges. However, this is caused by faulty switching equipment. The transmission plan deals with transmission quality under fault-free conditions. If attenuation is kept below a certain limit then the effects of noise not due to faults are rarely significant on inland calls. In other words, it is normally only when combined with severe attenuation that noise causes users to judge a call unsatisfactory. In the international telephone system, noise itself does become an important consideration. Echo is also a problem on international calls, and is even a problem within the larger national systems such as those of North America. In other types of switched system different types of impairment may have to be carefully controlled. In order to illustrate the principles involved in drawing up a transmission plan, we shall consider attenuation in a telephone system. Similar considerations apply to other types of impairment and other systems.

The hierarchy of exchanges forms the basis for the transmission plan. The transmission plan ascribes nominal values to the attenuation introduced by trunks on each category of trunk route. For the purposes of assessing the effects this has on users, a nominally worst-case local line is assumed to be used on all calls. (In practice some local lines produce more attenuation than this nominal worst-case.) A worst-case local line is the one used to decide on the area to be served by a local exchange. The geographical sizes of exchange areas are chosen to ensure that very few users have a local line whose attenuation exceeds this worst-case. This attenuation might be, typically, 10 dB, as measured at 1 600 Hz.

The results of putting a transmission plan into practice have to be predicted in order that the nominal values of the attenuation for each trunk route may be chosen. For example, consider the transmission plan illustrated in Figure 4.3. In this, all trunks between secondary and tertiary trunk exchanges introduce no attenuation at all. Singing does not occur because four-wire switching is used at these exchanges, that is the two directions of transmission are switched through the exchange separately. In order to prevent singing occurring over the complete four-wire switched part of the

127

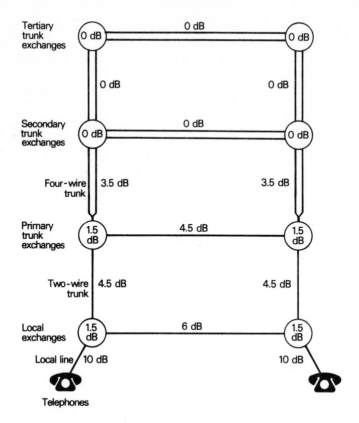

Figure 4.3 Example transmission plan. Four-wire switching is used in the secondary and tertiary trunk exchanges

connection, 7 dB attenuation is introduced between the two-wire points at the primary trunk exchanges. This is shown as 3·5 dB on each primary-secondary trunk route. No net attenuation is introduced by the four-wire exchanges; any attenuation due to switching equipment is compensated for by the first go amplifier on each link. Each two-wire exchange introduces 1·5 dB attenuation.

Consider a call using a backbone routing. The nominal total attenuation, adding the attenuations for each trunk and exchange, plus the worst-case local line attenuation is:

$$10 + 1{\cdot}5 + 4{\cdot}5 + 1{\cdot}5 + 3{\cdot}5 + 3{\cdot}5 + 1{\cdot}5 + 4{\cdot}5 + 1{\cdot}5 + 10 = 42{\cdot}0 \text{ dB}$$

However, on any given call using this routing, the total attenuation will not necessarily have this value. When each link is prepared for service, its attenuation will be set to the nominal value. After the link goes into service the values of many of its parameters, including attenuation, may drift, because of the ageing of components and changes in environmental conditions such as ambient temperature and humidity. The links will be tested periodically. If the values of any of the parameters have drifted outside a certain range, then the link is declared faulty, until it is repaired or adjusted by an engineer. Thus, links which are in service will have a range of parameters about their nominal values, but not outside the allowable limits according to maintenance policy. This is illustrated in Figure 4.4, which shows the probability of any given link, with a nominal attenuation of 4·5 dB, having a particular attenuation. This distribution, which is approximately a Gaussian distribution with the portions outside the range 2·5-6·5 dB cut away, is an adequate approximation to the distributions found in practice.

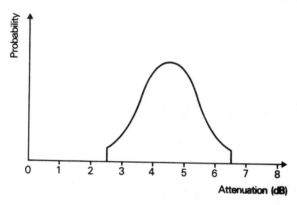

Figure 4.4 Probability distribution for attenuation produced by links with a nominal attenuation of 4.5 dB

If, in the above example of backbone routing, the 7 trunks are selected randomly from trunks whose attenuation probability distributions are similar to that of Figure 4.4, the result will be a probability distribution for the total attenuation something like that shown in Figure 4.5. This is very roughly a

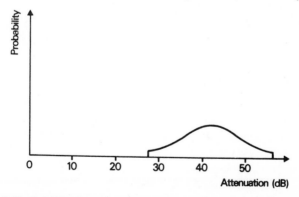

Figure 4.5 Probability distribution for attenuation of a transmission channel involving 7 links and having a nominal attentuation of 42 dB

Gaussian distribution, with a mean of 42·0 dB (the nominal total attenuation). It is much more spread out than the distribution for individual trunks. The standard deviation of a distribution representing a combination of values from n Gaussian distributions with standard deviations σ, is $(n^{1/2} \sigma)$, assuming no correlation between the n distributions.

Figure 4.5 represents the distribution of attenuations that users would get on backbone routing calls. There will be equivalent distributions for all categories of routing. Routings involving fewer trunk routes will have probability distributions for attenuation which are less spread out. To give a general picture of the probability of occurrence of a particular value of attenuation on all calls, the probability distributions for each category of routing are weighted according to the proportion of calls connected using each category of routing. (Most calls use routings involving only a few trunks; relatively few use backbone routings.) To translate this into a measure of what proportion of calls are judged unsatisfactory, it is necessary to have some experimental data on the opinions of users about attenuation. This would take the form of a graph of the proportion of users judging a call as unsatisfactory, versus the attenuation. The probability distributions for the occurrence of attenuation values are weighted according to this dissatisfaction distribution, and the result is integrated over all values of attenuation. This gives the proportion of all calls judged unsatisfactory because of attenuation. This is then a measure of the effectiveness of the proposed transmission plan. For example, the plan in the above example might be shown to lead to 5 per cent of calls being unsatisfactory. By a process of iteration, the plan can be adjusted to achieve the administration's target for quality of service, expressed in terms of the allowable proportion of unsatisfactory calls.

This gives a rough picture of the sort of procedures involved in deciding on a transmission plan. In practice it might be necessary to consider the combined effects of, for example, noise and attenuation, or echo and attenuation. A more detailed discussion of transmission plans, in particular international transmission plans drawn up by the CCITT, is given in References 40, 41 and 43.

4.5 RELIABILITY OF SERVICE

In order to ensure that the reliability of service is acceptable to users, it is necessary to define a set of reliability targets covering the various parts of a telecommunication system. This facilitates the optimisation of equipment design, that is making the equipment neither needlessly expensive nor unacceptably unreliable. Reliability targets thus express what is generally considered to be economically achievable.

The reliability target for each piece of equipment can be expressed in terms of the **probability of failure per unit time.** The unit of time normally used is one year. The reliability target for an exchange might be expressed as a probability of failure of 0·025 per annum. What this means, in terms of what is observable in practice, is that if there were, say, 4000 of these exchanges in the system then, on average, the number of exchanges which fail each year would be 4000 × 0·025 = 100. In this context, failures are assumed to occur randomly; probabilities and numbers of failures quoted in this way are only meaningful in terms of averages taken over many years, and for many similar pieces of equipment.

In deciding on reliability targets there are two factors which must be considered. First, there is the average time for which the failure lasts, often referred to as the **mean down-time** of the equipment in question. And secondly, there is the number of users affected by the failure. For example, a fault on a local line might remain for 24 hours before being repaired, but it affects only one user. However, a fault in an exchange might cause the exchange as a whole to fail for 5 hours, until it is repaired, and several thousand users might thus be affected. Some administrations have, in recent years, expressed their reliability targets as formulae, which contain the mean down-time and number of users affected as independent variables. Table 4.4

Table 4.4 *Examples of target probabilities of failure for various part of a telephone system*

Part of system	Target probability of failure per annum
Terminal	0·1
Local line	0·075
Local exchange switching equipment	0·01
Local exchange control equipment	0·01
Local exchange power supply	0·005

shows the sort of reliability targets that might be applied to various parts of a telephone system. Adding these gives the overall probability of service failure, for each telephone, as 0·2 per annum.

To find out whether a particular element of the system will meet its reliability target, an analysis of that element may be carried out when it is being designed. This analysis is based on values for the probability of failure per annum of components in the equipment. These values are derived from observations of the rate of occurrence of failures in large numbers of components in service in similar types of equipment. For example, the probability of failure per annum is typically 0·0003 for a reed relay, 0·0005 for a transistor, and 0·001 for an integrated circuit.

When thinking about equipment reliability it is important to distinguish between the probability of failure per unit time and the **fault rate.** The fault rate is the average number of faults occurring per unit time. The probability of failure is related to this, but not in a simple way. A single fault will not necessarily cause a complete failure of a particular element of the system. For

131

example, a single fault in an exchange may only affect one local line. It is only certain faults, and combinations of faults, which cause failures of the whole, or large parts of the exchange.

Many parts of a telecommunication system incorporate a certain amount of redundancy, that is there are a number of identical items of equipment performing a particular function, and several items may fail before the function as a whole becomes seriously affected. For example, in a strowger exchange, a fault in a selector will disable that selector, and may also affect a few other selectors. However, because each switching stage consists of a number of selectors, working more or less independently, the failure of a few of these does not cause the exchange as a whole to fail. The only effect noticed by users would be a slight worsening of the probability of call blocking. As it happens, strowger equipment has a high fault rate, as compared with other types of switching equipment, yet because of the high degree of redundancy inherent in its design, a strowger exchange as a whole is very reliable. It is largely the high reliability of strowger exchanges (typically having probabilities of failure per annum of the order of 0·02) which has led to the adoption of such strict reliability targets for new types of exchange by many administrations.

A topic which attracts a great deal of attention at present is the reliability of computers in s.p.c. exchanges. A simple computer incorporates no redundancy: a single fault will normally cause it to fail. So, although the electronic components used in computers have a low fault rate (a complete computer has a total fault rate of about 2 or 3 faults per annum), a simple computer is not satisfactory for controlling an exchange because each fault would cause a total exchange failure. Special fault-tolerant computers have to be used instead. These incorporate a certain amount of redundancy, and can therefore continue to operate even when a number of faults are present. Some of the most advanced fault-tolerant computers are claimed to have an overall probability of failure per annum as low as 0·01, assuming that faults are identified and repaired in an average of five hours. However, the derivation of figures such as this, on the basis of information about the reliability of the components used in the computer, involves a number of assumptions which are not entirely justified. A rigorous analysis would have to take account of possible deficiencies in the mechanisms which give the machine its fault-tolerance, and the complex interaction of the computer hardware (the machine itself) and the software (the programs which are stored in its memory). Some aspects of such analyses are discussed in Reference 33.

4.6 ECONOMIC ANALYSIS IN PLANNING

In planning additions to equipment in a continuously growing system there are usually a number of possible ways of meeting the demand for service. First, there may be several different versions of a particular element of the system to choose from, for example, two types of exchange equipment such as strowger and crossbar. Secondly, to keep the amount of equipment matched to the demand there may be a choice of times at which equipment can be provided. For example, it may be possible to provide extra equipment each year or, alternatively, equipment of twice the capacity every two years. If equipment is added every two years then half of it will not be fully used during the first year after provision. However, the costs of providing equipment before it is needed may be offset by the single two-yearly installation costing less than two yearly ones, owing to economies of scale.

In order to select one plan from several possible plans on economic grounds some method of economic analysis is needed. There are a number of different methods in use. Each has its advantages and deficiencies. One deficiency common to most methods of economic analysis is the inability of the analysis to take account of the long term results of each plan that cannot easily be described in terms of money. It may be obvious that the choice of a particular type of exchange equipment will have far-reaching effects on the system, and hence its overall cost, at some time in the future, but a description of these far-reaching effects is difficult to include in the analysis on which the choice of equipment is based. A second deficiency of some methods is that, even when substantial economic differences can be identified for a point in time many years ahead, the method of analysis is such as to diminish the apparent importance of events that are a number of years in the future, and thus overemphasise the more immediate cost differences. These comments are made here because there is sometimes a tendency to attach too much importance to the results of economic analyses and to ignore factors which the analyses, by definition, do not take into account.

The method of analysis which will be described here is the **present value of annual charges** method, or **p.v.a.c.** for short. It can be used to compare several possible plans with one another and to identify the one which is likely to incur the smallest overall cost. It does not provide any useful information about a particular plan considered by itself. The principle underlying p.v.a.c. (and several other methods) is that of **discounting.** If an expenditure of M units of money is incurred at a time n years in the future, this is considered to be equivalent to a smaller expenditure at the present time of a sum such that, if this smaller sum were invested in such a way as to earn interest at a rate r, it would grow to the sum M over n years. The smaller sum is said to be the sum M discounted back to the present time at interest rate r, and is given by: $M/(1 + r)^n$ where r is expressed as a fraction. (A 20 per cent interest rate would thus be expressed as $r = 0.20$.)

Each possible plan that is considered will commit the administration to three categories of expenditure, as follows:

(1) Capital expenditure, that is the money spent on providing the equipment.

133

(2) Continuing expenditure, that is the cost of having the equipment in service. This includes all running costs such as paying for supervisory and maintenance staff, spare parts, electricity, transport, and so on.

(3) Equipment replacement expenditure, that is the cost of replacing the equipment when it wears out. This might be typically every 15 to 30 years for electrical and electronic equipment, and 50 years or more for buildings.

Each time the equipment is replaced because it has worn out the cost incurred will be the cost of removing the old equipment, minus any scrap value it might have, plus the cost of the new equipment. The new equipment will generally cost more than the original equipment because of inflation, though it may cost less because of the use of more modern technology.

The principle of discounting is applied to all expenditures involved in each possible plan. The period over which these expenditures are considered may be a finite study period. Alternatively, and more simply, the expenditures can be assumed to go on indefinitely. This second approach will be used in the following example. Suppose that the expenditures involved for equipment provided in year n are:

(a) Capital expenditure, on providing the equipment, of C units of money in year n.

(b) Running costs of A units of money for each year from year n onwards.

(c) Replacement costs every 30 years, that is in years $(n + 30)$, $(n + 60)$, and so on. Owing to inflation the equipment will be assumed to increase in price by a factor b every 30 years, so the costs are bC for the first replacement, b^2C for the second, and so on.

Discounting all these back to the present time, assuming a constant interest rate r, gives a present value equivalent cost, P, of

$$P = \frac{C}{(1 + r)^n} + \sum_{k=1}^{k=\infty} \frac{A}{(1 + r)^{n+k}} + \sum_{m=1}^{m=\infty} \frac{b^m C}{(1 + r)^{n+30m}}$$

Summing the series gives

$$P = \frac{1}{(1 + r)^n} \left\{ C + \frac{A}{r} + \frac{bC}{(1 + r)^{30} - b} \right\}$$

Each plan will involve the provision of equipment in several different years in the planning period being considered. For example, there may be equipment provided in years 1, 3 and 5 in a five-year plan. Using the above expression for P, this would give values P_1, P_3 and P_5. The total present value equivalent cost for the plan would thus be $(P_1 + P_3 + P_5)$. For several possible plans, each specifying the provision of equipment in one or several of the years 1 to 5, the most economic one is the one with the lowest total present value equivalent cost.

A certain amount of caution is necessary in using methods of economic analysis such as p.v.a.c. Besides remembering the general limitations of the method mentioned earlier, it is important to check the sensitivity of the results to changes in parameters, such as the interest rate. If small changes in any parameter completely change the choice of plan, this indicates that the conclusions of the analysis are not to be relied upon. Factors not considered in the analysis should therefore be given careful consideration, since these may give a more reliable indication of which is the best choice. For example, if continually rising labour costs seem likely to make a substantial difference to the running costs of various types of exchange equipment, the possible results of this trend could be considered. This might indicate a preference for the use of equipment which seems to be more expensive under present conditions, but which requires less maintenance and may therefore prove to be more economic in the long run.

4.7 A SYSTEM OVERVIEW: THE BRITISH TELEPHONE SYSTEM

This chapter ends with an overview of the British telephone system. This overview is a 'snapshot' of a continuously evolving system taken in 1975.

4.7.1 Size and capacity
Britain has the third largest telephone system in the world, containing about 5 per cent of the world's telephones. In 1975 there were about 20 million telephones in Britain, that is about 0·35 telephones per head of population. The rate of growth of the system is about 8 per cent per annum, and if this continues there will be around 30 million telephones by 1980. In 1975 there were, on average, about 280 calls made per person per annum, making about 16000 million calls per annum in all. This means that, on average, a telephone call was made every 2 ms and, in the busiest periods of the day, calls were made at an average rate of one every 300 μs, with up to 1500000 conversations taking place at once.

4.7.2 International links and exchanges
In 1975 there were about 7500 trunks connecting the British telephone system to the rest of the world: 5000 provided by undersea cables, 1000 by communication satellite, and 1500 by line-of-sight microwave radio across the English Channel.

The total outgoing international traffic was about 3000 erlangs, and the incoming international traffic was about the same. Furthermore, the London transit centre is one of the six CT1s nominated by the CCITT, and it handled, on average, about 80 erlangs of transit traffic in 1975. (Because communication satellites are making it increasingly easy to provide small numbers of direct channels, the relative need for international transit switching is, in fact, diminishing.)

In addition to providing transit switching, London is also a significant patching-through point for international trunks, that is a number of channels in the various international transmission systems are not used by Britain at all, but are parts of transmission links for the use of other countries. The number of patched-through channels was about 950 in 1975.

All international trunks, except some from Eire which are connected direct to certain provincial trunk exchanges, are connected to the international gateway exchange in London. The gateway is a conglomeration of exchanges of different types and sizes (some crossbar equipment, some strowger) spread over many equipment rooms in different buildings in the City of London. The international trunks are brought up to London, from their points of arrival in Britain, by cable transmission systems and line-of-sight microwave radio systems.

The gateway exchange is fed with outgoing international calls from the inland telephone system by trunks which are connected either direct from group switching centres or via district and main switching centres. Calls incoming to telephones in Britain are fed into the inland trunk network from the gateway via the main trunk exchanges in London.

4.7.3 The exchange hierarchy

Exchanges in Britain are arranged into a hierarchy consisting of local exchanges, primary trunk exchanges, secondary trunk exchanges and tertiary trunk exchanges. The tertiary trunk exchanges are fully interconnected. In 1975 there were about 6200 local exchanges, 370 primary trunk exchanges (which are called group switching centres), 27 secondary trunk exchanges (which are called district switching centres), and 9 tertiary trunk exchanges (which are called main switching centres). There are a great many trunk routes between group switching centres, so that about 95 per cent of inter-area traffic can be connected without the use of district or main switching centres. This traffic is routed either over trunk routes between the two group switching centres, or via one intermediate group switching centre. District and main switching centres are used for the remaining 5 per cent of traffic.

Above group switching centre level in the hierarchy, 4-wire switching and multi-frequency digit signalling are used. This ensures that the attenuation on calls involving many trunk routes is acceptable and that the time taken to set up calls is not more than about 15 seconds.

The country is divided into about 650 numbering plan areas, each one containing, on average, 10 local exchanges. Because there are about 650 numbering plan areas, but only 370 group switching centres, many group switching centres have more than one numbering plan area to deal with; that is the local exchanges of several numbering plan areas may be served by the same group switching centre. The numbering plan area in which the group switching centre is situated is normally called the home area and the other numbering plan areas which it serves (if any) are called remote areas.

4.7.4 Exchanges in the trunk network

Group switching centres are sited in one of the larger towns in the area they serve. Most of the 370 group switching centres are closely integrated with the local exchange, or one of the local exchanges, serving the town in which they are sited, forming more or less a single exchange. In some towns the group switching centre is in a separate building on the outskirts of the town; this reduces the cost of trunks between the group switching centre and other trunk exchanges by eliminating the need to bring them into the centre of the town, where the provision of underground cables is expensive. In 1975, most of the 370 group switching centres were strowger exchanges, but about 40 were crossbar exchanges known as TXK1s. All 27 district switching centres and all 9 main switching centres were crossbar exchanges, known as TXK4s.

4.7.5 Signalling in the trunk network

In 1975 there were four principal signalling systems in use in the trunk network. These include two d.c. systems, known as loop-disconnect signalling (which is simply an adaptation of loop-disconnect local signalling), and signalling system d.c. No. 2 (or DC2 for short). DC2 is a version of loop-disconnect for audio links over 25 km long, and uses complete reversals of the current feed at the sending end, rather than breaks in current, for the digit pulses. The other two systems are in-band a.c. systems, called signalling system a.c. No. 9, or AC9 for short, and signalling system a.c. No. 11, or AC11. AC9 is a system which converts the signals of loop-disconnect into pulses of tone at one end, and back into loop-disconnect signals at the other. It uses a single tone for all signals. AC11 uses multi-frequency digit signals, but signals other than digit signals (often referred to as line signals) are the same as in AC9.

AC11 is used for trunks to and from district and main switching centres. (This helps to keep the time taken to set up multi-link trunk calls to a minimum.) AC11 is also used on a few trunk routes between group switching centres.

In 1975, there were a few p.c.m. transmission systems in the trunk network. These are ultimately to be used with common channel signalling between digital trunk exchanges. However, initially they had to be operated as self-contained transmission systems. The binary digital signalling sent over p.c.m. transmission systems is therefore converted to loop-disconnect signalling at the terminal equipment. A p.c.m. trunk is thus almost indistinguishable from a trunk using loop-disconnect, DC2, or AC9 signalling.

Table 4.5 shows the approximate proportions of the various signalling systems in use in the trunk network in 1975. The systems not specifically named are all obsolescent.

Table 4.5 *Proportions of various signalling systems in the trunk network, 1975*

Signalling system	Percentage used in trunk network
AC9	56
DC2	21
Loop-disconnect (including p.c.m.)	10
AC11	5
Others	8

4.7.6 Trunks in the trunk network

In 1975 there were over 6000 trunk routes between the various trunk exchanges in Britain. Some trunk exchanges had only a few routes connected to them, but the larger ones had several hundred. Each trunk route consisted of anything from four or five trunks to several thousand. The average was 50, but this was not necessarily the most common number. In all there were about 300000 trunks in the trunk network.

Each trunk may consist of any combination of channels from different transmission systems, connected end-to-end. In practice complete f.d.m. groups, or supergroups are patched through from one transmission system to another, and are only demodulated to individual channels where absolutely necessary. There is no way of giving a meaningful description of the combinations of transmission systems used on particular trunk routes. The actual combinations change from day to day as groups and supergroups are moved from one system to another in order that maintenance and installation work can be carried out. Although it is not meaningful to analyse the make-up of trunks in general, or even for a particular trunk route, the proportions of different types of transmission system in the trunk network can be analysed in terms of the number of circuit-kilometres provided by each type. Table 4.6 shows the relevant figures for the different transmission systems in use in 1975.

Table 4.6 *Circuit-kilometres of transmission capacity provided by various types of transmission system in the trunk network, 1975*

Transmission system	Circuit-kilometres in trunk network (millions)	Percentage of network
Coaxial cables (f.d.m.)	29	54
Microwave radio (f.d.m.)	10	18
Twisted pairs (audio)	10	18
Twisted pairs (f.d.m.)	3	6
Twisted pairs (p.c.m.)	2	4
Coaxial cables (p.c.m.)	Installation just started	—

These provided a total of 54 million circuit-kilometres which formed the 300 000 trunks in the trunk network.

4.7.7 Transmission systems in the trunk network

The use of twisted pairs to carry f.d.m. signals, providing 6 per cent of the trunk network transmission capacity in 1975, is obsolescent. In this type of transmission system one f.d.m. group is moved to the frequency band 12-60 kHz by single sideband modulation, using a 120 kHz carrier, with selection of the lower sideband. This is combined with a group in the normal 60-108 kHz band, so the two fit together to form a 12-108 kHz band containing 24 channels. Each system uses two 24-pair cables, one for one direction of transmission and one for the other, so as to minimise crosstalk between the two directions of transmission. Each twisted pair carries 24 channels in one direction, so each pair of cables provides 288 channels in all. The cables contain copper conductors of 1·27 mm diameter. Signals are attenuated by about 2·2 dB km^{-1} at 108 kHz, so for distances over 25 km repeaters giving up to 55 dB amplification are inserted at intervals of no more than 25 km.

The most common coaxial cable systems in 1975 were those providing 960 channels (16 supergroups) for every two coaxial tubes. In the earlier 960-channel (4 MHz) systems, coaxial tubes were used having a 2·6 mm inner conductor and 9·5 mm outer conductor (referred to as 2·6/9·5 mm coax for short). 2·6/9·5 mm coax produces about 6 dB km^{-1} attenuation at 4 MHz, so repeaters giving up to about 54 dB amplification are provided every 9 km. 2·9/9·5 mm coax is also used to carry 2700 channels (12 MHz) by reducing the repeater spacing to 4·5 km.

In the later systems a small bore, 1·2/4·4 mm coax, was used. This produces an attenuation of about 13 dB km^{-1} at 4 MHz. Repeaters are spaced 4 km apart for a 960-channel system and 2 km for a 2700-channel system. The latter was the standard arrangement for new systems in 1975. Two coaxial tubes are needed for each system (one for each direction of transmission), and a cable containing several tubes may serve several systems. In 1975, cables were in general use containing 2, 4, 6, 8 and 12 tubes. A 60 MHz (10800 channel) system was introduced into the system in 1975. This uses the large bore coax, with 1·5 km repeater spacing. At the same time an 18-tube cable was introduced, so that each cable could carry nine 10800-channel systems, giving almost 100000 channels per cable.

The older repeaters in the network contained valves (vacuum tubes) and needed high power supply voltages. A 50 Hz, 1000 volt supply was fed through the cables using the two centre conductors of the coax. The later repeaters, using transistors, need only 20 volts d.c. For these repeaters, the two conductors carry $\pm$ 250 volts. The current is limited to 50 mA and the repeaters draw power in series. Small buildings were used to house the older repeaters but more modern repeaters are placed in small manholes by the roadside. Cables are buried in ceramic, metal, or asbestos ducts, similar to drainage pipes. Each duct can hold several cables and new cables are drawn into the ducts as required and jointed together in small manholes.

Table 4.7 shows the proportions of different types of coax systems in use in Britain in 1975, according to the circuit-kilometres they provided.

Table 4.7 *Proportions of different types of coaxial cable transmission systems, 1975*

Coax systems	Percentage of total coax circuit-kilometres
Large bore, 4 MHz	10
Large bore, 12 MHz	34
Small bore, 4 MHz	40
Small bore, 12 MHz	16

Audio circuits use 0·90 mm copper conductors or, more recently, 0·63 mm copper conductors. Cables with aluminium conductors are also used. These have thicker conductors because of the higher resistivity of aluminium. Some audio circuits are loaded with 88 mH coils at intervals of 1·83 km (2000 yards) so as to improve the transmission characteristics of the circuit in the 300-3400 Hz speech band. For circuits which constitute the complete trunk (that is they do not form the tail end of an f.d.m. channel), two-wire circuits without amplification are used up to about 25 km. Two-wire circuits are used with n.i.c. repeaters up to 40 km. In order to introduce amplification, many circuits are converted to four-wire form by means of hybrids at each end. Circuits are always four-wire if they are joined onto the tail end of an f.d.m. channel.

The earlier p.c.m. systems provided 24 channels on two twisted pairs taken from an ordinary audio cable. The regenerative repeaters, spaced 1·83 km apart, replaced the loading coils used on audio circuits. These repeaters were powered from a ± 75 volt d.c. supply, fed between the two pairs. Ordinary audio cables were taken over to provide more circuits on existing routes without making it necessary to provide new cables. These earlier 24 channel p.c.m. systems had a line bit-rate of 1·536 Mbits s⁻¹. (The CCITT standard 30+2 channel systems, operating at 2·048 Mbits s⁻¹, were introduced into the system after 1975.) Coax p.c.m. systems, working at 120 Mbits s⁻¹ (1920 channels) were introduced in 1975.

Microwave radio, although providing only 18 per cent of the network capacity in 1975, makes itself very conspicuous in the shape of the Post Office Tower in central London and its 120 little brothers scattered around the country. The rapid growth in the use of microwave radio in the late 1960s was stimulated by the need for a large number of 5·5 MHz television channels to carry colour television broadcasts to transmitting stations around the country. The towers are spaced roughly 40 km apart and every transmission system involves terminal equipment at each end of the system and a repeater at each intermediate tower.

Between any two towers, up to six systems can operate. The centre frequencies of the bands used are those recommended by the CCIR: 1·8 GHz, 2·1 GHz, 4·0 GHz, 6·175 GHz, 6·76 GHz and 11·19 GHz, though the systems using the two lowest bands are obsolescent. The space between two towers is effectively used up once a system operating in each band has been installed. This provides enough capacity for over 100 000 channels, the lower bands being able to carry fewer channels than the higher ones. However, not all of this capacity can actually be used for traffic. Because of fading and equipment failures, at least one high capacity channel (960 or 1 800 speech channels) is used as a protection channel in each system. Where a system carries television and telephone channels, the telephone hypergroups have precedence over television channels for the spare. So, if a telephone hypergroup and a television channel both fail, and there is only one spare hypergroup, then the television channel is lost.

Because only a certain number of channels can be carried between any two points using microwave radio, there is a limit to the capacity of the microwave network. Although that limit had not been reached in 1975, in the long run the microwave network will provide a decreasing proportion of the transmission capacity of the trunk network.

4.7.8 Local exchange networks

Centred on each group switching centre is a local exchange network. This consists of all the local exchanges in one or several numbering plan areas, the trunks linking these to the group switching centre, and a few direct trunks between certain local exchanges where the traffic passing between them is significant. The group switching centre acts as the access point to the trunk network for inter-area calls.

The group switching centre provides all the translation facilities for inter-area calls; the local exchange simply repeats the dialled digits to the group switching centre, by means of inter-exchange signalling, and the group switching centre takes over the setting up of the rest of the call. The reason for doing this is that it would be uneconomic to store all translations for inter-area calls at each local exchange. Because the routing from the group switching centre onwards is independent of which local exchange the call originates from, only one set of translations is needed at the group switching centre; if these were stored in each local exchange the information would be needlessly replicated. The main local exchange of the town containing the group switching centre often shares a building with the group switching centre. Large towns usually have more than one local exchange.

London, Birmingham, Edinburgh, Glasgow, Liverpool and Manchester are known as **director areas.** The local networks of these areas are arranged slightly differently from other areas. They have a considerable number of local exchanges (almost 400 in London) and the function of the group switching centre is split between a number of separate trunk exchanges and tandem exchanges within each metropolitan area. London is divided into seven sectors which are treated almost as separate numbering plan areas for certain types of call.

Because most trunks within local networks are less than 25 km in length, twisted pairs are used with loop-disconnect signalling for almost all trunks. The pairs and loading arrangements used for the trunks are the same as for audio trunks in the trunk network. Cables contain up to about 1000 pairs. Wherever possible more than one cable is used for each trunk route so that a cable failure does not completely disable the route. Also, different cables serving the same trunk route follow geographically separate paths.

The total number of local network trunk routes was about 14000 in 1975. The total number of trunks was about 850000, so the average route size was about 60 trunks.

4.7.9 Local exchanges

There are three types of switching equipment used in local exchanges in Britain: strowger, crossbar, and reed relay. The strowger exchanges are normally categorised as follows:

(a) Director exchanges: strowger exchanges incorporating register-translators, and built up piecemeal to the required size.
(b) Non-director exchanges: strowger exchanges without register-translators, and built up piecemeal to the required size.
(c) Unit automatic exchanges (UAXs): strowger exchanges installed in fixed-sized units (without register-translators).

Director exchanges are used in the director areas: London, Birmingham, Edinburgh, Glasgow, Liverpool and Manchester. They make it possible to implement local number dialling, whereby within-area calls are made by dialling just the local number. Non-director exchanges are used in most other towns and large villages in Britain. UAXs are used in small towns and villages with less than 2000 telephones. There are several slightly different versions of the UAX, the most common being code-named the U13.

There were two types of crossbar local exchange in use in Britain in 1975 called the TXK1 and the TXK3. The TXK1 replaces non-director exchanges, that is local exchanges with more than 2000 telephones. The TXK3 replaces director exchanges. Both the TXK1 and TXK3 provide extensive register-translator facilities so that almost any numbering scheme can be implemented. The main reed-relay exchange in use in 1975 was the TXE2. This replaces UAXs and provides no register-translator facilities. However, it can distinguish between an own-exchange call and a call via the group switching centre, without a special code having to be dialled by the user. With non-director exchanges and UAXs it is normally necessary to make the user dial a 9 before certain numbers to give access to the group switching centre.

So, if the group switching centre provides all the necessary facilities for local number dialling, these are automatically extended to TXE2 local exchanges. Local number dialling facilities, provided by the group switching centre, are similarly extended to TXK1 exchanges, but these can be used for local number dialling without the aid of the group switching centre.

Table 4.8 shows the approximate numbers of different types of local exchanges in Britain in 1975.

Table 4.8 *Local exchanges in the British telephone system, 1975*

Local exchange type	Approximate number in use	Percentage of telephones connected to type of exchange
Director	550	36
Non-director	1350	46
UAX	3400	10
TXK1 and TXK3	300	4
TXE2	600	4

1976 saw the first public service of a larger reed-relay exchange called the TXE4. This can act as a combined group switching centre and local exchange for up to 40000 lines. The TXE4 forms one of the bases of the modernisation plan for the British telephone system.

4.7.10 Local lines

Local lines consist of pairs of conductors. These are normally twisted pairs in cables. The normal local distribution scheme used in Britain is for a number of cables (main cables), containing up to 4000 pairs, to go from the exchange to a number of metal cabinets throughout the exchange area. These cabinets are cross-connection points. From the cabinets a number of secondary cables (branch cables) radiate to smaller cross-connection boxes, housed in asbestos pillars. These are normally on street corners and serve one or more streets. On large housing estates the pillars are sometimes omitted and the street cables taken direct from a cabinet. The tertiary cables (distribution cables) radiate from the pillars (or direct from a cabinet) and run down streets under the pavement. Individual pairs are taken off at distribution points either on wooden poles or under the pavement. If poles are used (the cheaper but less elegant method) then overhead wires are strung across the street to the houses. Underground cable, usually containing two pairs (in case a second telephone is needed) is an alternative method for linking houses to the street cable; the cost of this is normally subsidised by the builder of the houses on new estates. Distribution cables are fed directly into office buildings and other premises needing a number of telephones. Main cables and branch cables are buried in ducts like trunk cables, but distribution cables are normally directly buried in the ground under the pavements.

The combinations of cable used for local lines are determined by the need to keep the attenuation of speech below about 10 dB (as measured at 1600 Hz), and the line resistance below 1000 Ω to ensure satisfactory local signalling, while minimising the thicknesses of conductor used. The four types of conductor widely used are 0·40 mm, 0·50 mm, 0·63 mm and 0·90 mm diameter. The thickest wire is used nearest to the telephone and the thinnest nearest to the exchange. Telephones near the exchange have local lines consisting of the thinnest gauge (and therefore cheapest) wires only, but those further from the exchange generally have local lines which start off thin (thus sharing main cables with telephones near the exchange) and become thicker away from the exchange. The longest local lines use thicker gauge wire all the way. The 10 dB nominal limit on attenuation restricts local lines to about 11 km length using 0·90 mm conductors. This wire produces about 0·95 dB km^{-1} attenuation and has a resistance of 55 Ω km^{-1}, so an 11 km local line is well within the resistance limit. In practice it is sometimes necessary to exceed 11 km length, and so exceed 10 dB attenuation. Local lines are not normally loaded.

4.7.11 Telephone terminals

The basic telephone circuit introduced in Britain in 1959 is shown in Figure 4.6. Telephone terminals based on this circuit are called 700-type telephones. (This simply means that they are known by three-digit codes beginning with 7.)

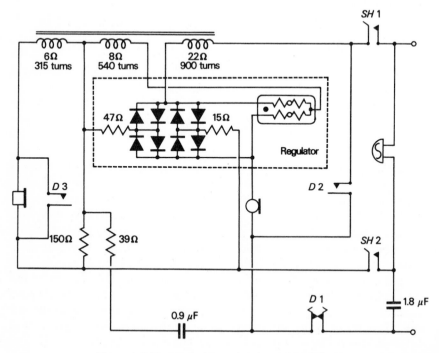

Figure 4.6 Basic telephone circuit used in Britain.

In 1975 the standard table-top instrument was the 746. There are numerous other telephones for different purposes, including pay-phones, but all use the same basic circuit.

The basic 700-type circuit is similar to the telephone circuit described in Chapter 3. A special feature is the automatic regulator, containing two resistors, eight diodes, and a two-filament device similar to a light bulb. The resistance of the filaments increases with the current flowing through them. The regulator introduces attenuation which is greater the shorter the local line, so that the transmission performance of the telephone plus local line is roughly independent of the local line length; the shorter the local line, the greater the current flowing through the regulator filaments, and hence the greater the attenuation introduced.

Other substantially different terminals which can be used in Britain include loudspeaking telephones, facsimile machines, telemetry machines, and data modems. Local lines are also connected to private branch exchanges of various kinds. In this situation the local line effectively becomes a trunk between the branch exchange and local exchange. However, the local signalling remains the same as if an ordinary telephone were at the user's end of the local line, except where the branch exchange has direct dialling in. In this case the branch exchange is linked to the local exchange by a trunk as if it were another public exchange. In 1975 about 4 million telephones were connected to the system through private branch exchanges, instead of direct to a local line. Apart from the convenience to the organisations concerned, of being able to make calls between telephones on a branch exchange by dialling only a few digits, and without incurring a charge for the use of the public system, the use of branch exchanges increases the efficiency of the public system by concentrating traffic onto fewer local lines than the number of telephones; local lines connected to branch exchanges each carry 0·5 erlang or more.

4.7.12 Shared service

To further economise on the use of local lines, two residential users with low calling rates can be made to share one local line. This arrangement is called shared service. The bells in the telephones are connected between one wire of the local line and earth, so that each telephone can have its bell rung independently by the connection of ringing current through one wire with earth return. On outgoing calls the user is identified, for billing purposes, by a signal at the start of the call. This is generated by the user pressing a button on the telephone before dialling. The button momentarily earths one wire of the local line and the exchange can identify which user is calling by detecting which of the wires has been earthed. Only one user can make or receive a call at a time. About 30 per cent of residential telephones (20 per cent of all telephones) had shared service in 1975.

A new form of local line sharing, introduced in 1975, effectively gives each telephone its own local line. This is the $1+1$ carrier system. The $1+1$ carrier system is a simple f.d.m. arrangement in which one telephone uses the line in the normal manner, but the other has its speech and signalling amplitude modulated on to a 40 kHz carrier from the telephone to exchange, and a 64 kHz carrier in the other direction. The presence of the carrier does not affect the operation of the other telephone. The power needed for the speech circuit and ringing the bell in the carried telephone is supplied by a rechargeable battery, which is trickle-charged from the exchange via the local line when no call is in progress.

Chapter 5

Monochrome television systems

INTRODUCTION

The second part of this book, that is Chapters 5 and 6, deals with some aspects of national television systems which use radio transmission. Television systems have been chosen because they are quite unlike the telephone and other systems considered in the first part, and this makes it possible to introduce ideas and techniques which are very different from the ones treated in previous chapters. An overall description of television systems will not be given. Instead, we shall concentrate mainly on receivers and the constraints they place on the systems, because the ideas we believe to have the strongest claim for inclusion in a book of this length arise when one looks at the choices which have to be made in deciding on the principal technical features of receivers.

Receivers have to be cheap enough for the domestic market and capable of producing pictures which are acceptable to the majority of viewers. It is these two factors, one to do with economics and the other to do with human visual perception, which make receivers central to the design of television systems. The economics of broadcast television and telephone systems are radically different. The transmission channel for television, the atmosphere in which radio waves are propagated, is available at no monetary cost. There is a social cost in using a scarce commodity—radio broadcast bandwidth—for one purpose as opposed to another. This may have a bearing on the overall price we are prepared to pay for having television systems. But once the decision to have a system has been taken, the cost of the transmission channel can be treated as nil, whereas in telephone systems, the transmission channels, which include exchanges as well as lines, account for a major part of the total cost. The whole cost of television systems lies in the terminals. The total cost of transmitter terminals, of which there are only a few to serve the entire country, is relatively small and, in general, the major cost of a television system lies in its receivers.

One primary consideration in making the various choices which have led to the current systems has therefore been minimisation of receiver cost. The cost of a receiver increases with the quality of the picture it is capable of producing. Minimising receiver cost implies designing for the minimum acceptable quality. Our assessment of picture quality depends on the nature of human visual perception which is therefore a starting point for receiver, and hence system, design.

This chapter deals with monochrome (black and white) television in which picture information is present as a variation of brightness over the surface of a luminous screen. A discussion of the way we measure brightness in terms of perception data (Section 5.1) is followed by a very brief description of a television system and of how a television picture is built up on the receiver screen (Sections 5.2 to 5.4). Perception data is then used to establish the main picture parameters and the required bandwidth of the vision signal (Sections 5.5 and 5.6). The use of synchronising pulses is described in Section 5.7. The type of modulation chosen to transfer the sound and picture signals onto an r.f. carrier, the forms of distortion involved and the receiver characteristics they entail are discussed in Sections 5.8 to 5.10. The overall form of the transmitted signal is described in Section 5.10. In Section 5.11 the structure of a receiver is outlined in terms of the various functions and elements discussed in the previous sections. Finally, the way in which the non-linear response of picture tubes is dealt with is described in Section 5.12.

5.1 LIGHT AND PHOTOMETRY

Before dealing with television systems in any detail, you need to be clear about some properties of light, their physical significance, and the way they are expressed in engineering terms. It is not sufficient to know about purely physical aspects of light, because the design of television systems relies heavily on certain physiological features of the human perception of visual information. Among these are the facts that the sensitivity of the eye varies with colour and that we can perceive much more brightness information than colour information.

The relevant properties of light and of visual perception will be dealt with in two stages: the ideas relevant to monochrome television will be introduced in this section and they will be extended for use in dealing with colour television in Chapter 6, Section 6.2.

Light consists of electromagnetic waves, just like radio waves, but at a much higher frequency, that is much smaller wavelength. The eye responds to wavelengths from about 380 to 780 nm (1 nm $= 10^{-9}$ m), that is frequencies of about 420 to 790 THz (1 THz $= 10^{12}$ Hz). Most people recognise different wavelengths as the different colours of the visible spectrum shown in Figure 5.1. The sensitivity of the eye varies with wavelength. Figure 5.2 shows the relative response of the eye when viewing a monochromatic (single frequency) source which emits a constant power but whose wavelength can be varied over the whole visible range. This curve, which is analogous to the frequency response of a filter, is known as the **visibility function,** the **photopic response** or the $\bar{y}$ **graph.** It was obtained by averaging the results obtained from a large number of human observers and has been internationally agreed as a standard. The response of any individual viewer is likely to depart somewhat from the visibility function but it provides

148

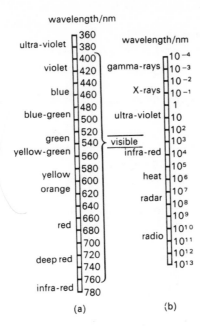

(a)   (b)

Figure 5.1 (a) The visible spectrum. (b) The electromagnetic spectrum

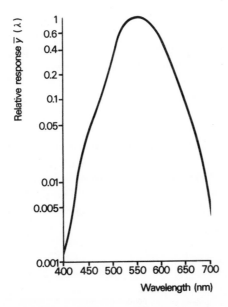

Figure 5.2 Average visual response to electromagnetic radiation. The *power* response $\bar{y}(\lambda)$ relative to the response at the wavelength giving maximum sensitivity is plotted against wavelength

a working objective standard for human visual response. You can see from Figure 5.2 that a source of red light at 680 nm would need to be about a hundred times stronger (in watts of luminous power) to have the same visual effect as a source of yellow-green light at 550 nm for which the eye has maximum sensitivity.

The visibility function provides a link between purely physical quantities and their perception by human beings. It is similar in this respect to the psophometric weighting curve of Figure 3.1 which represents the relative effect of the frequency component of electrical noise fluctuations on the human intelligibility of speech messages.

The measurement of light energy regardless of human perception factors forms part of the science of *radiometry*. Measurements taking human perception into account are the province of *photometry*.

Photometry involves many units and related quantities which often cause a good deal of confusion. This is partly due to the need to differentiate between measurements of light in three situations: light emitted by direct sources such as light bulbs; light incident on illuminated surfaces; and light emitted indirectly by reflection from illuminated surfaces which only reflect part of the light reaching them. If you are interested in the various units and their interrelations, they are discussed in References 62 and 70. Fortunately we need only consider photometric quantities associated with direct sources when dealing with television. This is because we will only be concerned with the scene being televised and its image on the cathode ray tube. The image is made up of luminous traces on the screen of the tube, that is direct sources. The various parts of the scene are normally only luminous because they reflect sunlight or the studio lights, but we can treat the scene as if it were made up of direct sources, because we will not be concerned here with the subtle art of the lighting engineer.

The photometric quantities we will consider are flux, luminance, and luminous intensity.

To understand what is meant by photometric flux consider a light source whose power density spectrum is represented by the graph of $P(\lambda)$ against wavelength λ in Figure 5.3. Since we are dealing with a power density spectrum, the quantity $P(\lambda)d\lambda$ represents the power *in watts* radiated by the source in a narrow band $d\lambda$ at wavelength λ. The total power (energy per second) radiated by the source in the visible range is the area under the curve which is given by

$$\text{total power radiated} = \int_{\lambda\,=\,380\text{ nm}}^{\lambda\,=\,780\text{ nm}} P(\lambda)d\lambda \quad \text{in watts}$$

This is the power flowing from the source, in other words, the flux of electromagnetic energy. However, it is not a photometric quantity because it

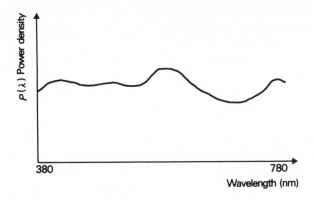

Figure 5.3 Power density spectrum of a light source (note that the power density is plotted against wavelength, rather than frequency, which is the variable usually chosen when dealing with electrical signals)

takes no account of the variation of human visual sensitivity with wavelength. This can be done by using the visibility function of Figure 5.2. At each wavelength λ, we need to multiply $P(\lambda)$ by $\bar{y}(\lambda)$, the ordinate in Figure 5.2. Figure 5.2 is normalised; it represents the relative response with the maximum value arbitrarily chosen to be unity. A multiplicative factor K_m is chosen to give a convenient photometric unit of **flux**. This unit is called the **lumen** (lm) and we have

$$\text{luminous flux } \varphi = K_m \int_{380}^{780} \bar{y}(\lambda) \, P(\lambda) d\lambda \tag{5.1}$$

with $K_m = 680$ lumens per watt (lm W^{-1}). The reason for choosing 680 is that it makes some SI photometric units nearly equal to the corresponding units (such as candle power) which were previously in common use.

Luminous flux is the total photometric power generated by a source in all directions. In dealing with television, we are usually concerned with **luminance** which provides an objective measure of how 'bright' the source is perceived to be. The luminance of a light source is the luminous flux per unit solid angle per unit projected area in the direction from which the source is being viewed. The luminance of a light source is therefore $\varphi/(\omega a \cos \theta)$ where φ is the luminous flux generated by the source into solid angle ω, a is the area of the source, and θ is the angle between the normal to the surface of the source and the direction along which it is being viewed; so that $a \cos \theta$ is the projected area of the source.

Luminance is usually expressed in **candelas per square metre** (cd m^{-2}), the

151

candela being the basic SI unit of luminous intensity. The **luminous intensity** of a source is the flux generated by the source into a unit solid angle, hence

luminous intensity (in candelas) = flux per steradian

and

luminance (flux per steradian
per square metre)   = candelas per square metre

(The steradian is the SI unit for solid angles. The surface of a sphere subtends 4π steradians at any point inside its surface.)

Hence the luminous intensity of a source (in candelas) is its flux per steradian, and its luminance (flux per steradian per square metre) is therefore the number of candelas per square metre it radiates.

The property of a scene which is reproduced by a black and white photograph is, ideally, the luminance of its various parts. Similarly, in television, the luminance of the scene is reproduced on the receiver screen by ensuring that the luminance of each element of the screen is approximately proportional to the luminance of the corresponding element in the scene.

5.2 THE AUDIO AND VIDEO SIGNALS

Figure 5.4 represents a simple television system. The camera and microphone convert the visual and aural information in the scene into electrical signals which are modulated on to a carrier and amplified in the transmitter so as to produce radio frequency signals which can be broadcast as electromagnetic radio waves from the antenna. The receiver converts these radio waves back to a lower frequency electrical signal which is used to produce a picture on the screen of a cathode ray tube and sound coming from a loudspeaker.

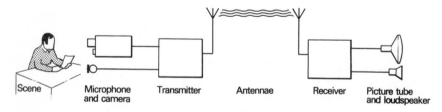

| Scene | Microphone and camera | Transmitter | Antennae | Receiver | Picture tube and loudspeaker |

Figure 5.4 The basic elements of a television system

Figure 5.4 is a specific example of a very simple system. Several cameras may be used or the electrical signals may initially be obtained from video tape recorders, from telecine machines or slide scanners. These last two convert films or photographic slides into appropriate signals. In the case of test patterns, signals may be generated electronically by the use of appropriate oscillators. The signals produced by these various sources are usually processed in a video control centre. The transmission links between the video centre and the main transmitter are normally either coaxial cables or microwave radio systems.

The transmission of sound information in the system of Figure 5.4 involves converting the air pressure fluctuations due to sound waves at the microphone into an electrical signal which, after modulation, transmission, demodulation and amplification, is used to reproduce the appropriate air pressure fluctuations by means of the loudspeaker. In an ideal distortionless system the pressure produced by the loudspeaker is directly proportional to the microphone pressure, though there is, of course, a delay between the corresponding inputs and outputs due to the time taken by the message signal to travel through the system. The electrical signal corresponding to the sound is called the **audio signal.** It is the electrical analogue of the sound information being transmitted through the system. The electrical audio signal represents the sound pressure fluctuations in a direct and simple way.

The situation is not so simple as far as the visual information and the corresponding electrical signal are concerned. The picture perceived by the viewer is a two dimensional display consisting of grey areas, the extreme values of grey being termed black and white. This picture is built from the motion of a luminous spot over the surface of the cathode ray tube. At any instant the spot occupies a small area of the screen whose brightness (luminance) determines what shade of grey is perceived for that area. The instantaneous value of the electrical vision signal, which is called **video signal,** therefore only carries information about one small area in the picture and about the corresponding region in the scene on which the television camera is focused. It is not a direct electrical analogue of the complete visual information as it is perceived by the viewer at that instant.

5.3 THE SCANNING PROCESS

When a television camera is focused on a scene, an image of the scene is formed on a light sensitive surface in the camera so as to produce an electrical charge pattern. The amount of charge at each point depends on the luminance of the corresponding element of the scene. This charge pattern is converted into a current by means of an electron beam which is made to scan over it. This current flows through the camera load resistor and the voltage across this resistor is the video output signal. The detailed operation of cameras will not be described here (see References 89 to 96 if you wish to find out more about this). All you need to appreciate is that the video signal at any instant represents the luminance of a small portion of the scene, and that the electron beam inside the camera is scanned in a systematic way so as to ensure that the video signal represents the luminance of each element of the scene in turn.

The scanning of the electron beam is achieved by simultaneous horizontal and vertical deflections of the beam as in a cathode ray oscilloscope. It can be produced electrostatically or magnetically. Magnetic deflection coils are commonly used in cameras to produce the horizontal and vertical motions of the beam. A simple scanning pattern is shown in Figure 5.5. The horizontal motion is called the **line scan** and the vertical motion the **field scan.** In the line scan the left to right motion of the beam (A to B in Figure 5.5(a)) is much slower than the return motion (B to C). The return is known as the **line flyback.** Similarly, the downward motion of the field scan is much slower than the upward **field flyback.** Figure 5.5 is drawn so that there are eight

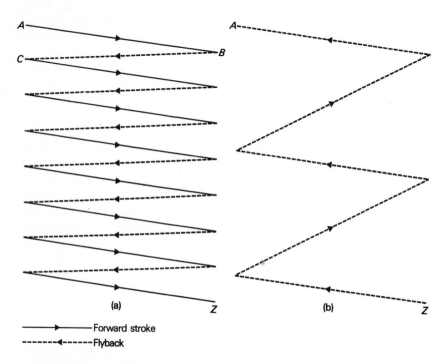

Forward stroke

Flyback

Figure 5.5 A simple scanning pattern. The trace is blanked out during flyback. There are eight active lines per field and two (blanked) lines during the field flyback period

complete line scans during one downward stroke of the field scan, and two complete line scans during the field flyback. These are relatively small numbers, chosen to illustrate the scanning method. Actual numbers for a broadcast system are discussed in Sections 5.5 and 5.6.

The pattern traced out by the electron beam in the camera tube is called a **raster.** Two pairs of magnetic deflection coils are used to make the electron beam follow a similar raster on the screen of the receiver cathode ray tube. The visual message information received is processed so as to make the beam current of the cathode ray tube vary with the video signal in the transmitter. The luminance of the light spot formed on the receiver screen therefore depends on the magnitude of this signal which, in turn, depends on the luminance of a small part of the scene.

In order to produce a satisfactory picture of the scene, the camera and receiver scans must be synchronised so that, for instance, when the receiver beam is scanning point A of Figure 5.5, the beam current is proportional to the video signal for the point corresponding to A in the camera image. The way in which this can be done will be discussed in greater detail in Section 5.7. It involves the transmission of **synchronising pulses** (or **sync pulses,** as they are often called) to control the start of each line and field scan. These pulses are transmitted during the flyback periods. This is possible because the video signal carries no luminance information during flyback periods and the beam can be **blanked.** That is to say, the beam current can be so reduced that the light spot is no longer visible.

The visual message signal therefore consists of a combination of the video signal and the sync pulses. It is called the **composite video signal.**

The scanning process for one complete picture takes, typically, 40 ms. That is 25 complete pictures are generated each second. The light produced on the screen is due to the fluorescence caused by a beam of electrons. Fluorescence ceases when the beam moves away. However, the light does not stop immediately owing to another effect, phosphorescence, which produces an afterglow. It dies away gradually taking typically 4 ms to drop to 5 per cent of its initial value. Thus at any instant the brightness of about 36/40 (90 per cent) of the picture is less than 5 per cent of its peak value and a photograph of the screen taken with a sufficiently fast exposure, say 0·1 ms, would reveal less than 10 per cent of the television picture. A photograph taken with an exposure of 40 ms would reveal the complete picture as the whole screen would be scanned during that time. The restricted response of the human eye causes the picture perceived by an observer to be somewhat like the photograph obtained with a 40 ms exposure. Although, physically, there is a light spot of variable intensity scanning rapidly over the area of the screen, a human observer perceives a picture filling the whole of the screen all the time.

The **line period** is the time taken for the spot to move over the whole path ABC in Figure 5.5(a). The number of lines in a complete picture is an important system parameter and some of the factors affecting its choice are discussed in Sections 5.5 and 5.6. It is usual, when specifying this number, to include the blanked lines which are swept out during the field flyback (Figure 5.5(b)). These do not carry luminance information. The lines which do carry luminance information are called **active lines.**

5.4 TRANSMITTER AND RECEIVER: BASIC ELEMENTS

A very simplified block diagram of the basic elements of a television transmitter is shown in Figure 5.6(a). The sync pulses which are added to the

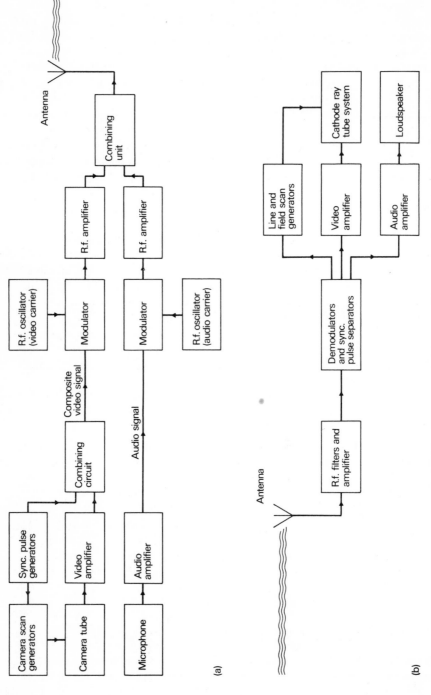

Figure 5.6 Block diagrams of (a) a television transmitter; (b) a television receiver

video signal are also used to synchronise the time bases which produce the camera tube scan.

Two radio frequency (r.f.) carriers are used. One is modulated with the composite video signal and the other with the audio signal. The modulated r.f. carriers are amplified and combined before being fed to the transmitter antenna. A transmitter would have many other elements, including amplifiers to bring the various signals to appropriate levels before combining them, filters to ensure that the signals do not occupy a larger bandwidth than the system requires, and various correction circuits to compensate for known imperfections of the system. Some of these circuits will be discussed later on.

A similarly simplified block diagram of a receiver is shown in Figure 5.6(b). The filter and amplifier connected to the antenna represent the tuned circuits used to select the appropriate television transmission channel. The r.f. signal is processed in the demodulators which have been drawn as a single box but consist of various stages of demodulation and amplification, including intermediate frequency (i.f.) amplification, which will be discussed later. The outputs of the demodulators consist of three signals:

(1) the audio signal which is amplified and fed to the loudspeaker;

(2) the video signal, corresponding to the camera output, which is used to control the current in the electron beam of the cathode ray tube;

(3) the sync pulses which are separated from the composite video signal and used to control two sawtooth oscillators which provide the scanning waveforms. The outputs of these oscillators are amplified and used to drive two pairs of coils located on the neck of the cathode ray tube so as to produce a magnetic field which deflects the electron beam, both horizontally and vertically, to make it scan over the whole face of the tube. Some receivers with small tubes use electrostatic, instead of magnetic, deflection.

5.5 PICTURE PARAMETERS

The raster of Figure 5.5(a) only has eight lines. If a raster with a much larger and more realistic number of lines had been drawn, the lines would have been much closer together and the detailed motion of the spot would have been much harder to follow. This is just the situation aimed at in high definition broadcast television. The larger is the number of lines, the more uniformly does the raster cover the screen and the less visible are the individual lines which make up the picture. The amount of detail which the picture can convey increases with the number of lines, but the bandwidth needed to transmit the picture information also increases. This is an instance of the rule that the signal bandwidth needed to convey information increases with the amount of information to be conveyed. A compromise therefore has to be reached since bandwidth is at a premium because there are many telecommunication systems competing for a portion of the limited radio

157

frequency spectrum; and because the complexity and cost of amplifiers and other system elements increases considerably with bandwidth.

The principal parameters of a television picture, which include picture definition, line and field periods and number of fields per complete picture, play a considerable part in determining the details of the electrical signals transmitted and processed by the system. They therefore have an important bearing on many of the technical choices made in setting up the system, and they form a good starting point in the study of these choices.

We will now consider how the requirement to produce a picture with adequate definition, while using a minimum signal bandwidth, can be combined with certain features of visual perception so as to provide guide lines for the choice of picture parameters.

5.5.1 Visual acuity

This is a measure of the amount of detail which can be perceived visually. Since transmitting more detail than is necessary would result in a waste of bandwidth a measure of visual acuity is essential.

Visual acuity is measured in terms of the angle subtended at the eye by the centres of two identical small objects which can just be separately identified.

A high degree of acuity implies the ability to distinguish objects which are very close together, that is objects which subtend a small angle α at the observer's eye (Figure 5.7). For this reason visual acuity, V, is defined as the reciprocal of the angle

$$V = \frac{1}{\alpha}$$

The SI unit for V is the reciprocal radian.

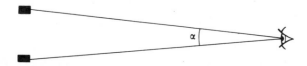

Figure 5.7 Visual acuity is the reciprocal of the angle α subtended at the eye by two small objects which can just be resolved

Acuity has been measured for a large number of people. It depends on many factors such as the shape of the objects, the contrast between the objects and their background, and the intensity of the light to which the eye is adapted. Tests in which these factors were chosen to be appropriate for the conditions which normally obtain when viewing a television picture have led to a generally accepted value of 3 400 reciprocal radians (that is approximately one reciprocal minute) for the design of black and white television systems.

5.5.2 Number of lines in a picture

Visual acuity decreases with decreasing contrast so that the sharpest picture which can be resolved by a viewer assumed to have a visual acuity of $1/\alpha$ is a pattern of black and white lines subtending an angle α at the viewer's eye. Figure 5.8 represents such a pattern of horizontal lines on a television screen height H. If the viewer is distance d from the screen, the separation between

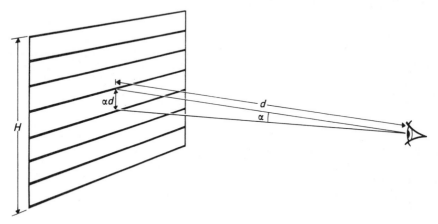

Figure 5.8 Horizontal line pattern on a television screen.

consecutive lines on the screen is very nearly equal to αd, since $\sin \alpha \simeq \alpha$ for small α. The total number of lines on the screen is therefore $H/(\alpha d)$ or, since $V = 1/\alpha$, the number of lines which can just be resolved is equal to HV/d.

Now, a viewer is likely to sit closer to a small screen than a large one. User trials carried out in a number of countries have led to the conclusion that, on average, viewers tend to sit at a distance approximately equal to eight times the screen height. This is taken as a design figure, so that using this value for d/H and 3400 reciprocal radians for V, we conclude that a maximum of $3400/8 = 425$ lines can be resolved in a television picture under accepted standard conditions.

This does not take into account the fact that the television picture itself is made up of lines which occupy fixed positions on the screen. This means that the number of lines which needs to be transmitted is greater than the number of lines which can be resolved, as is shown by the following argument.

Consider Figure 5.9. It shows two extreme situations which can arise when trying to display on the screen a set of horizontal dark bars whose width and spacing are equal to the spacing between the raster lines. The figure represents vertical sections of the original pattern and of the screen. The sections of the lines on the screen are shown as circles to indicate how the spot luminance falls off *between* lines relative to its value *on* the lines (relative because the absolute luminance will depend on the magnitude of the video signal at each particular instant during the scan).

In Figure 5.9(a) the position of the dark bars coincides with that of alternate lines, with the result that alternate lines are dark and bright and the pattern on the screen looks somewhat like the original bars which can therefore be resolved by the viewer. In Figure 5.9(b) the bar pattern has been

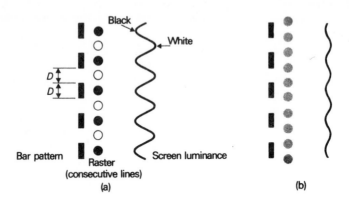

Figure 5.9 (a) The transmitted bar pattern is coincident with alternate raster lines. (b) The transmitted bar pattern has the same spacing as in (a), but is displaced by one half-line spacing

shifted by half the inter-line distance. Now each line is equally bright and no bar pattern can be resolved.

The viewer will just be able to resolve this pattern for some intermediate position between the two extreme cases of Figure 5.9(a) and (b). Thus even though he may be able to resolve the bars whose width and spacing is D, he may not be able to resolve them if they are presented as a picture having a line structure whose inter-line distance is D. Statistical tests show that the average viewer can always resolve a bar picture if the inter-line spacing is KD, where K is called the **Kell factor.** Values for K ranging from 0·53 to 0·85 have been reported by various experimenters, but a value of 0·7 is generally accepted for television systems.

Putting things the other way round, if the average viewer can just resolve 425 lines on a surface the size of a television screen at normal viewing distance, then the raster used to build up a similar line pattern will need to have $1/K$ times this number of lines, that is $425/0·7 = 607$ lines.

This figure of 607 provides us with an initial estimate of the number of lines required in a television picture. The actual value chosen in a system depends on what compromise is made between picture detail and economy of bandwidth. Various public systems use 405, 525, 625 (the CCIR standard) and 819 lines. These numbers include lines traced out during the field flyback period which do not form part of the picture; this is discussed in Section 5.6. The 405- and 819-line systems will, eventually, be replaced by 625-line systems. We will return to the relation between video signal

160

bandwidth and the number of lines in a picture, but the next step is to consider what determines the number of separate pictures which need to be shown in unit time and, from this, the field period and frequency.

5.5.3 The flicker effect

The scanning process in television, together with the relatively rapid decay in spot luminance after the electron beam has passed by, cause each area of the screen to produce short light pulses with much longer intervening dark periods. If the interval between successive pictures is too long, that is if for a given pulse duration (which depends on the properties of the screen phosphor), the pulse frequency is too small, various parts of the picture appear to flicker on and off. This is called the **flicker effect.** The frequency below which the flicker effect becomes apparent varies from observer to observer. It also depends on the average luminance of the background and on the shape of the light pulse. Experimental results, averaged over a large number of observers, for a range of background luminance and for a spot luminance pulse shape typical of television pictures, are shown in Figure 5.10, together with details of the experimental pulse shape.

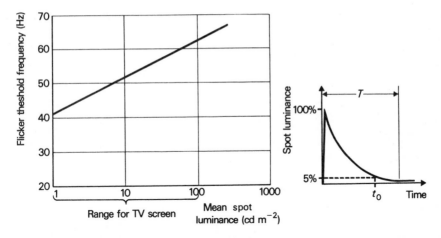

Figure 5.10 Frequency below which flicker becomes apparent, plotted as a function of mean spot luminance. The results are averaged over many experiments using a spot with a 4 µs decay time to 5 per cent of peak luminance. (Picture tubes used for monochrome television have 5 per cent decay times ranging from 3 to 5 µs)

5.5.4 Interlaced scanning: picture and field frequencies

It appears from Figure 5.10 that a picture frequency of slightly more than 60 per second is required if flicker is to be avoided for the range of spot luminance used in television receivers. Now the bandwidth needed to transmit n pictures per second is proportional to n, so that the video signal bandwidth is proportional to the picture frequency and any acceptable means of keeping this frequency down is worth using. It might appear at first sight that the picture and field frequencies must be equal, so that, once the field frequency has been chosen, there is no more to be said. There is, however, an advantage to be gained by arranging for the picture frequency to be an integral sub-multiple of the field frequency by the use of **interlaced scanning.**

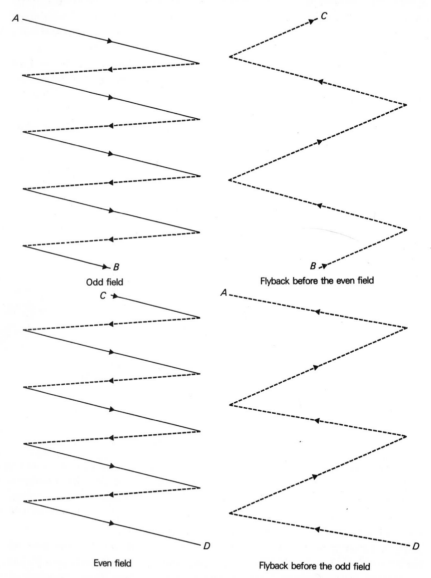

Figure 5.11 Interlaced scanning with two fields per complete picture. The blanked traces during line and field flyback are shown as dashed lines

Figure 5.11 shows the line pattern for the type of interlaced scanning in which there are two fields per complete picture. This is used in all broadcast systems.

The spot scans from A to B during the first field. The field flyback then takes it from B to C, from where it moves to D during the second field. The flyback finally takes it back to A when the whole process starts again.

The combined effect of the two fields is shown in Figure 5.12. Adjacent lines which belong to alternate fields are uniformly spaced provided that points A and C are accurately located. This is a matter of ensuring adequate accuracy for the timing and amplitude of the scanning waveforms.

The complete picture, that is the total number of lines, is produced at half the field frequency. However, flicker is perceived as affecting separate small areas and not the picture as a whole. These small areas extend over distances corresponding to a few line spacings and therefore appear to be illuminated at the field frequency. It is this frequency which is relevant to the flicker effect.

A picture containing N lines can be produced by interlacing two consecutive fields, each containing $N/2$ lines. If the field frequency is F, there are $\frac{1}{2}NF$ lines per second and the signal needed to carry the luminance information for these lines can occupy only half the bandwidth required for the NF lines of a picture having the same definition, i.e. the same number of lines, but without interlacing.

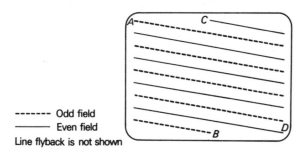

Figure 5.12 **Raster** consisting of two interlaced fields. Line and field flyback are not shown in this figure

One might gain even more in bandwidth by interlacing 3, 4 or more consecutive fields, but this higher order interlacing has not proved satisfactory because the resulting decrease in the frequency with which each individual line is illuminated brings about a different flicker effect. This makes the whole picture appear to vibrate.

The experimental results of Figure 5.10 suggest that a field frequency of slightly over 60 Hz is required if flicker is to be avoided. A frequency of 60 Hz is used in North America and some other countries. However, a field frequency close to the supply mains frequency has also been chosen in countries with 50 Hz mains for a number of reasons. Among these is the fact that there is always some residual mains frequency ripple on the voltages of the d.c. supplies in a.c. mains operated receivers. This produces a modulation effect on the vision signal. The degree to which this effect is noticeable depends critically on the difference between the ripple and the field frequency. It is most noticeable when this difference is between 7 and 10 Hz.

Another reason for choosing a field frequency close to the supply mains frequency is that stray magnetic fields, due to mains transformers, may cause the electron beam in the picture tube to be deflected. This effect is least noticeable when the frequency of the magnetic field is the same as the scanning frequency, because the resulting interference then takes the form of a static distortion.

The supply mains frequency is 50 Hz in Europe and 60 Hz in America. It can be seen from Figure 5.10 that, from the point of view of flicker, 60 Hz is much better than 50 Hz. Even so, the advantages of having the field scan at supply frequency prevail in Europe. American receivers, because of their higher field frequency, can be operated at higher luminance levels before flicker becomes objectionable. Remember that Figure 5.10 represents average results used in selecting system parameters. Individual viewers can establish their own compromise between maximum luminance and flicker by adjusting the brightness or contrast control on their receivers. The contrast control sets the gain in the vision channel, and thus determines the maximum luminance of the picture for a given brightness setting.

5.6 THE BANDWIDTH OF THE VIDEO SIGNAL

Having dealt with some of the consequences of various visual perception effects we are in a position to estimate the video bandwidth needed in order to produce a high definition television picture. The video bandwidth depends on the fineness of detail that can be seen in the picture, both horizontally and vertically. That is it depends on the horizontal and vertical resolutions of the picture.

Figure 5.13 represents a chequer board pattern displayed on a television screen. We will consider how fine a pattern can be resolved visually, assuming that we require equal vertical and horizontal resolution. A system which could display more detail than a viewer can perceive would simply be wasting bandwidth. We already know that we need about 600 horizontal lines to achieve adequate vertical resolution. We will take the number of horizontal lines, N, in the picture (including the lines which are generated during the field flyback period), as a starting point. This is convenient because N is an important and frequently quoted system parameter. Indeed in the United Kingdom it is usual to refer to the v.h.f. and u.h.f. networks as '405 lines' and '625 lines' which are their respective N values.

The pattern of Figure 5.13 is made up of horizontal lines which form the raster. Each horizontal line consists of alternate light and dark elements.

Equal horizontal and vertical resolution implies that, if there are N horizontal lines in the picture, there must be AN alternate black and white elements, called **picture elements,** along each horizontal line, where A is the **aspect ratio** of the picture, that is the width to height ratio. A is chosen to be 4/3 in all national broadcast systems. One reason for this choice is that it is a standard value used for films.

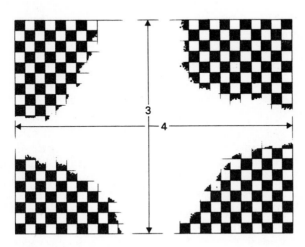

Figure 5.13 Chequer board pattern used to determine maximum resolution of a television display

Our initial estimate of AN picture elements per line must be modified for two reasons:

(1) The number N is the *total* number of line scans generated by the horizontal scan generator, or line scan generator, as it is called. Some of these line scans take place during the field flyback period when the beam is blanked, thus the number of active lines displayed on the screen is significantly less than N. It is equal to a_1N where a_1 is a number less than 1, typically between 0·91 and 0·94 for the main public broadcast systems. Note that the desired figure of 607 horizontal lines arrived at in Section 5.5.2 was $a_1 N$ and not N. In the British 625-line system $a_1 = 0·92$, which gives 0·92 × 625 = 575 active lines. This is close to our value of 607 lines which was based on appropriate statistical estimates of visual acuity, viewing distance, and the Kell factor. The agreement is closer for the 625-line system than for earlier (405, 525, and 819) systems, as can be seen from the active line numbers for these systems which are tabulated in the Appendix.

(2) The argument of Section 5.5.2 which led to the introduction of the Kell factor does not apply to the horizontal resolution because the spot moves continuously in the horizontal direction. As explained in Section 5.5.2 N is

chosen to be greater than the number of horizontal lines which the system can resolve vertically by a factor $1/K$. So, if we start with N, we need to multiply it by a factor K when considering horizontal resolution.

Thus the maximum number, N_v say, of distinguishable elements in one line of the picture is

$$N_v = a_1 K A N \tag{5.2}$$

The frequency band of the luminance (video) signal extends from d.c. (corresponding to a uniform blank picture which is white, black or any intermediate shade of grey, depending on the d.c. level) to the maximum frequency corresponding to a picture with the maximum possible detail. Each horizontal scan line of such a picture would consist of $\frac{1}{2}N_v$ bright segments separated by $\frac{1}{2}N_v$ dark ones. This pattern of luminance along a line can be produced, as indicated in Figure 5.14, by a sinusoidal video signal whose frequency, f, is equal to $N_v/2T$ where T is the time for one line scan, excluding the flyback time. The bright and dark segments are produced respectively by the signal maxima and minima.

A sinusoidal signal is chosen because, from a frequency spectrum point of view, it is the simplest, lowest frequency signal which will produce a visible pattern of N_v elements. A square wave signal would produce a sharper picture but it would contain higher frequencies. We are concerned with the lowest frequency signal which can provide the required detail and, in any case, we are dealing only with statistical estimates of visual perception parameters.

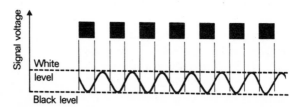

Figure 5.14 A sinusoidal signal used to produce a regular bar pattern on the screen of a picture tube. The contrast between light and dark regions has been exaggerated. However, the non-linear response of the tube (see Section 5.12) will lead to a sharper contrast than that produced by a pure sinusoidal variation of luminance

Hence we have:

$$f = N_v/2T \tag{5.3}$$

and N_v is given by Equation (5.2), so in order to obtain an estimate for f, we need only work out T.

If the field frequency is F then with N lines per picture and two fields per picture, the period for one line is $2/(FN)$. Now some of this time is taken up by line flyback and T, which is the line period excluding flyback, can be expressed as

$$T = a_2 \frac{2}{FN}$$

where a_2 is a number slightly less than 1 which represents the proportion of a line period during which the beam is not blanked. The values specified for a_2 in the major public broadcast systems range from 0.805 to 0.826. In other words, the line flyback takes up between 17.4 per cent and 19.5 per cent of a line period.

Using this expression for T together with Equations (5.2) and (5.3) we obtain the following estimate for the maximum required video signal frequency:

$$f = \frac{AK a_1 FN^2}{4a_2} \tag{5.4}$$

Since the video signal frequency can go down to d.c., the estimated maximum useful frequency, f, is also an estimate of the bandwidth required for the signal.

We have only dealt so far with the *luminance* signal. The complete video signal also includes the various synchronising pulses which are required to control the scan generators and ensure accurate interlace. These pulses occur when the beam is blanked, that is when no luminance information is being transmitted. It turns out that the bandwidth required for the luminance signal is more than adequate for the synchronising pulses. Equation (5.4) therefore gives an estimate for the bandwidth of the complete video signal. In the British 625-line system $A = 4/3$, $K = 0.7$, $a_1 = 0.922$, $a_2 = 0.813$, $F = 50$ Hz and $N = 625$; so Equation (5.4) gives $f \simeq 5.17$ MHz. The actual value for the system is 5.5 MHz. It differs slightly from the calculated value, which is based on a number of estimates for visual perception data, and on the assumption that the spacing which can be resolved between picture elements is equal to the spatial wavelength of a checker pattern produced by a sinusoidal signal. However, in so far as the estimates are valid, the result indicates that the horizontal and vertical resolutions are approximately equal for the British system. Similar calculations for the 405-line system using the data of Table A.1 in the Appendix show that the horizontal resolution is better than the vertical for this system.

Bandwidths ranging from 4.2 MHz (in Argentina) to 6 MHz (in France, Russia, and parts of Africa) have been used for 625-line systems. All these have the same vertical resolution, because they have the same number of lines. The difference lies in the horizontal resolutions which are proportional to the bandwidth. Thus the horizontal resolution is less than the vertical for the 4·2 MHz bandwidth, and greater for the 6 MHz bandwidth. (The horizontal resolution is also inversely proportional to the field frequency, but all the 625-line systems use 50 Hz.)

5.7 THE VIDEO WAVEFORM

Figure 5.15(a) represents part of a video waveform for consecutive lines. The signal is coded so that the range of levels between 0 and 33 per cent of the maximum level corresponds to sync pulses, and the remainder of the range corresponds to luminance information, with black corresponding to 33 per cent and white to 100 per cent.

For the British 625-line system, the field frequency is 50 Hz, so there are 25 pictures per second and $25 \times 625 + 15625$ lines per second. The **line frequency** is therefore 15 625 and the **line period** is the reciprocal of this, which is exactly 64 μs.

Figure 5.15(a) provides a convenient way of representing the video signal, but it can appear in different, though equivalent, forms in various parts of the system. It can, for instance, be shifted in level and passed through an inverting amplifier to give the waveform of Figure 5.15(b). Either type of waveform can be used to modulate the r.f. carrier in the transmitter. Using the waveform of Figure 5.15(a) gives what is known as **positive modulation,** using the waveform of Figure 5.15(b) gives **negative modulation.**

With negative modulation, the sync pulses shown in Figure 5.15(b) occur during the periods when the spot is blanked and when the magnitude of the signal exceeds 77 per cent of its peak value. Notice that the luminance increases with decreasing voltage. Negative modulation is used in America, Japan, and in all European systems, except for the French 625- and 819-line systems and for the obsolescent British 405-line system. These use positive modulation in which the spot luminance increases with the video signal and the sync pulses correspond to low values of the signal. Impulsive interference produces black spots on the screen with negative modulation and white ones with positive. White spots are more noticeable and cause more irritation to the viewer, mainly because spot size increases with brightness, so that a given amount of interference produces a larger spot with positive than with negative modulation. Impulsive interference can have more effect on the sync pulses, and hence on picture synchronisation, for negative than for positive modulation. However, this does not present a problem in practice because the synchronisation of the scan generator usually relies on the average timing of several pulses. This is known as **flywheel synchronisation.** It considerably reduces the perturbing effects of single interference pulses.

If the amplitude of the composite video signal at the transmitter becomes

168

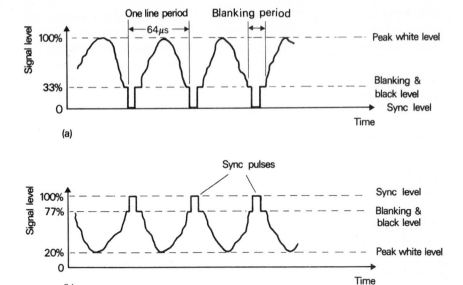

Figure 5.15 Composite video waveform for consecutive line periods for the 625-line system (a) with peak white corresponding to maximum signal level (positive modulation); (b) with the sync level corresponding to maximum signal level (negative modulation)

too large it will cause overloading and distortion in the modulator or in the r.f. amplifier, and hence in the transmitted signal. With positive modulation this affects the highlights, that is the brightest parts of the picture; with negative modulation it affects the sync pulses, so that distortion due to overloading does not produce a visible effect, so long as synchronisation is not impaired, which is normally the case for overloads of less than 30 per cent.

Figure 5.15(b), which represents the modulation used with the British 625-line system, can be obtained from Figure 5.15(a) by reflecting the vertical scale about the 100 per cent line and compressing it so that it covers 80 per cent of the range, instead of the 100 per cent of Figure 5.15(a). Thus a percentage level V in Figure 5.15(a) becomes $(100-80V/100)$ per cent in Figure 5.15(b). With this form of signal, maximum spot luminance, that is the **peak white level,** corresponds to approximately 20 per cent of the maximum signal level; and minimum spot luminance, that is the **black level,** corresponds to approximately 77 per cent of the maximum level. Notice that, with negative modulation, the video signal does not extend down to zero. Low signal levels would be affected by small fluctuations due to noise. The choice of a minimum level of 20 per cent prevents this from happening.

169

The spot cannot be seen if the magnitude of the signal rises above the black level, which is therefore also the **blanking level.** The blanking and black levels do not coincide in all systems—as can be seen from Table A.1 in the Appendix. They sometimes differ by a small percentage of the maximum signal level. This ensures that small errors in signal level, amounting to less than the difference between the blanking and black levels do not produce a visible effect on the screen during flyback.

It can also be seen from Table A.1 that different systems use different percentages for the black and peak white levels.

In the remainder of this chapter diagrams of the form of Figure 5.15(a) with zero sync level, 33 per cent black level and 100 per cent peak white level, will be used to represent the video signal for the British 625-line system.

5.7.1 The scan generators

The pulses shown in Figure 5.15 which lie between the blanking level and the line which is labelled **sync level** are used to synchronise the line scan generator. Before discussing the form of these pulses and the more complex field synchronisation pulses, it is necessary to be **clear** about the function of the scan generators which they control.

The magnetic fields produced by the line and field scan coils deflect the electron beam. This makes the luminous spot on the screen move as shown in Figures 5.11 and 5.12. The line scan on its own produces a horizontal motion to and fro across the screen, much like the time base of an ordinary laboratory oscilloscope. The field scan on its own produces a similar vertical motion at a much slower rate than the line scan. In both cases the forward stroke (left to right for the line scan, top to bottom for the field scan) takes place much more slowly than the corresponding return, or flyback, stroke.

If a given luminance level in a particular region of the scene being televised is to produce a corresponding level in the displayed picture, independently of where the region is located, the motion of the luminous spot during the forward strokes must relate exactly to that of the scanning spot in the camera. Cameras are designed to make the scanning spot move with constant velocity, and hence to sample equal portions of the scene in equal times. Thus, ideally, the receiver spot should also move with constant velocity

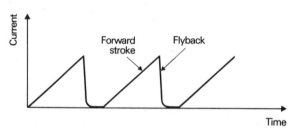

Figure 5.16 Linear sawtooth scanning waveform

during the forward strokes. If the spot deflection were exactly proportional to the deflection coil current, the output current of the scan generators would need to have a sawtooth waveform as shown in Figure 5.16 and the forward stroke would need to be linear. The exact form of the flyback is not critical, provided it is sufficiently rapid. The timing of the scanning waveforms must be accurately maintained from cycle to cycle so that the output of the receiver

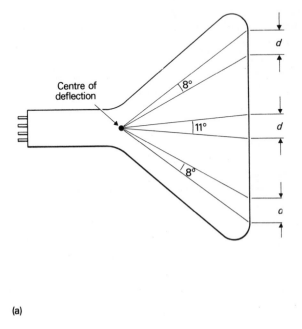

(a)

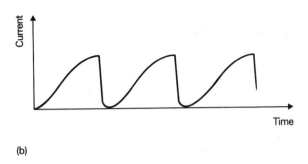

(b)

Figure 5.17 (a) Different deflection angles are needed to produce the same spot displacement, d, in different parts of the screen.
(b) Non-linear sawtooth waveform used to compensate for the effect illustrated in (a)

171

scan generators can be precisely controlled at every instant. This is done by means of sharp synchronising pulses which are applied to the generator once per cycle. These pulses, which usually initiate the flyback, ensure that the waveforms are in synchronism with the camera scan.

The scanning coils produce an approximately uniform magnetic field for which the *angular* deflection of the electron beam is directly proportional to the coil current. However, the *distance* moved by the spot across the screen is directly proportional to the current only if the screen surface forms part of a sphere whose centre coincides with the centre of deflection of the beam. Tubes are made with approximately flat faces and, as indicated in Figure 5.17(a), this means that a given deflection corresponds to a larger angular deflection at the centre of the screen than at the edges. The output circuits of scanning generators are designed to give the kind of scanning current waveform which is needed for this type of screen (Figure 5.17(b)).

5.7.2 The line synchronisation pulses
A more detailed diagram than Figure 5.15 of the waveform for one complete line period is shown in Figure 5.18. The time reference for synchronising the line scan generator is obtained from the leading edge of the sync pulse. (The leading edge is at the start of the pulse that is on the left-hand side of Figure 5.18.) The pulse is passed through a differentiating circuit and the rapid rate of change of voltage with time at the leading edge produces a sharp spike

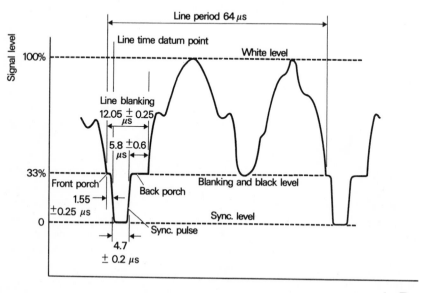

Figure 5.18 Composite video waveform for one line period

which is used to control the start of the flyback in the line scan generator. The leading edge is preceded by a short period at blanking level called the **front porch** which is provided to ensure that the video signal has time to return to the black level before the synchronising pulse starts, irrespective of the video signal level at the extreme end of the line. This is necessary because of the finite rise and fall time of the circuits through which the signals are passed. If the video signal corresponds to a high luminance at the end of a line, it will take a finite time to reach the black level. Without the front porch this would cause a delay in the start of the sync pulse, and hence an error in synchronisation. The front porch allows a settling period before the start of the sync pulse. The **back porch** extends the blanking period to an overall length of 12 μs. This is chosen to ensure that the spot is blanked during the whole of the flyback.

5.7.3 The field synchronisation pulses

The pulses involved in the field synchronisation for the British 625-line system are shown in Figure 5.19. They are much more complex than the line sync pulses, mainly because they must ensure accurate interlace. The various factors leading to the form of the pulse trains shown in Figure 5.19 will now be considered in turn.

Referring back to Figure 5.11, you can see that there are two types of field: even fields which start at C, half way through a line, and odd fields which start at A, at the beginning of a line. Thus blanking, which is applied during field flyback, ceases half way through a line at the start of an even field, and at the beginning of a line at the start of an odd field. The same thing applies in Figure 5.19. The first visible part of the even fields starts half way through line 18 and the first visible part of the odd fields starts at the beginning of line 331. The lines are numbered from an arbitrary datum line shown in the figure. This datum line coincides with the start of the field broad pulses which will be discussed later.

Looking at the upper trace of Figure 5.19, you can see that the picture is blanked during the period corresponding to the first 17½ lines of the even fields and the last 2½ lines of the odd fields, that is 20 lines in all. Similarly, you can see from the lower trace that the picture is blanked for 20 lines during the period at the end of even fields and the start of odd ones. The field blanking periods extend over 20 lines to allow for the field flyback which takes several line periods, and for a delay between the end of the flyback and the start of the field scan, so that the field scan generator can settle down after the flyback. The field blanking extends over an exact number of line periods determined at the transmitter. Twenty lines are shown in Figure 5.19. Between 18 and 22 are specified for the 625 monochrome system and 25 for the British colour system.

If the field scan is to be precisely related to the line scan, the line scan generator must continue to operate regularly throughout the field blanking

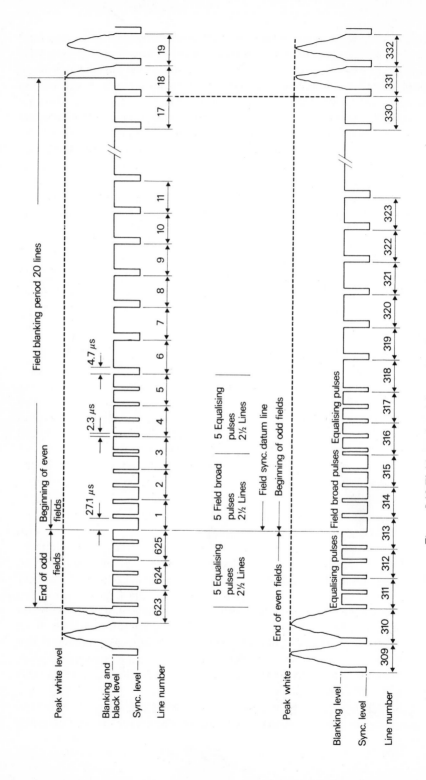

Figure 5.19 Field synchronising pulses for odd and even fields

period. For this reason the pulse train is designed to continue providing line sync pulses throughout this period. This is done by ensuring that there is a pulse leading edge at the beginning of each line period. There are some extra pulses, among the ones labelled equalising pulses and field broad pulses (which are discussed below), but they do not affect the operation of the line scan generator. They occur half way through a line and this is a time when the line scan generator is insensitive to sync pulses.

Pulses which are intended to synchronise the field scan generator must be electrically distinguishable from those which only synchronise the line scan generator. The distinction could be made in terms of pulse height or pulse width. Pulse width (duration) is chosen because this is more economical of transmitter power. The wider pulses are used to synchronise the field scan. They are often called **field broad pulses.** Their duration is $27 \cdot 1 \ \mu s$ compared with $4 \cdot 7 \ \mu s$ for the line pulses. A conceptual block diagram of part of a receiver is shown in Figure 5.20. The complete train of sync pulses is separated out from the composite video signal by the sync separator and fed to circuits which discriminate between line and sync pulses by their duration, and process them so as to produce pulses of appropriate shape to synchronise the two scan generators.

The line sync generator produces a sharp pulse to coincide with the arrival of the leading edge of each line sync pulse. The field sync pulse generator produces one pulse for each train of five field broad pulses.

Various types of circuit are used to discriminate between line and field sync pulses and to convert these pulses into a suitable form for synchronising the line and field scan generator. The basic principles of one way in which this can be done are illustrated in Figure 5.21. Two resistor-capacitor circuits are used. The first (Figure 5.21(a)) acts as a differentiator. Its time constant, RC, is short compared with the duration of the input pulses. The output waveform, which is obtained across the resistor, consists of sharp negative and positive impulses approximating in form to the time derivative of the input voltage.

The second circuit (Figure 5.21(b)) acts as an integrator. Its time constant is larger than the input pulse duration (though only slightly larger in practice). The output waveform, which is obtained across the capacitor, approximates to the integral of the input.

Figure 5.21(c) shows the result of separately passing the sync pulses from the upper waveform of Figure 5.19 through a diffentiator and through an integrator. The output of the differentiator contains negative and positive impulses. The positive impulses are removed by a clipping circuit. The negative impulses are used to synchronise the line scan generator. There is an impulse at the start of each line. There are also impulses half way through some lines but these do not affect the synchronisation process. You can verify that an impulse corresponding to the start of each line is also obtained if the

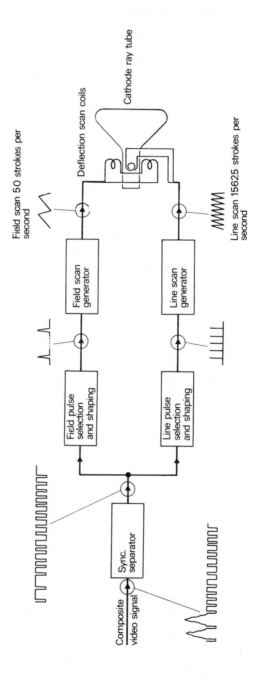

Figure 5.20 Block diagram of synchronisation and scanning circuits in a television receiver

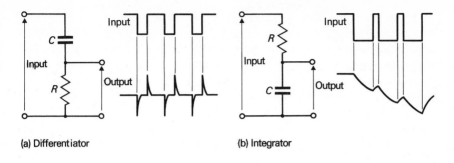

(a) Differentiator

(b) Integrator

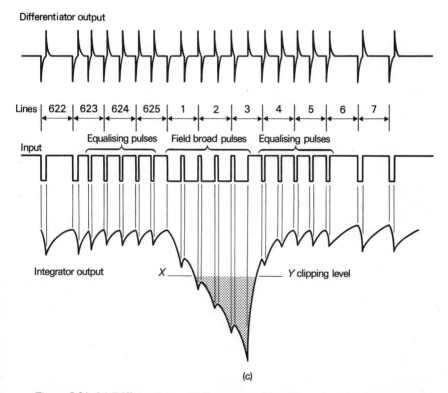

(c)

Figure 5.21 (a) Differentiator. (b) Integrator. (c) Differentiator and integrator outputs produced by equalising and field broad pulses, using the pulse train corresponding to the upper waveform of Figure 5.19

sync pulses from the lower waveform of Figure 5.19 are passed through a differentiator.

177

The output of the integrator consists of a set of negative going pulses. The integration process causes the composite output pulse corresponding to the field broad pulses to be much larger than the other output pulses. The integrator output is processed by a clipping circuit so that only voltages below the clipping level XY are passed on to the field scan generator synchronising circuit; which therefore receives one pulse per field. This pulse is the shaded area below the clipping level in Figure 5.21(c).

The accuracy of the interlacing of consecutive fields depends critically on the timing of the field scan. Any shift by more than about 5 per cent of a line period ($0·016$ per cent of a field period) will cause consecutive picture lines to appear to bunch in pairs. The timing of the field scan depends on the exact shape of the pulses obtained from the integrator (essentially because the scan generator synchronising circuit is sensitive to the energy in these pulses and the energy of a pulse depends on its area). The shape of these pulses must be the same for even and odd fields and the equalising pulses are used to ensure this.

It can be seen from Figure 5.19 that, without equalising pulses, there would, at the start of an even field, be one line period between the first field broad pulse and the previous pulse (the line sync pulse at the start of line 625). Whereas, at the start of an odd field, there would only be half a line period between the first field broad pulse and the previous pulse, (the line sync pulse at the start of line 313). It can also be seen from Figure 5.19 that a similar discrepancy would occur at the end of each set of field broad pulses for consecutive fields.

The equalising pulses ensure that there is a similar pulse pattern immediately before and after the field broad pulses for both even and odd fields. This ensures that the pulse shape below the clipping level XY of Figure 5.21(c) is the same for both types of field. You can verify this by sketching the integrator output for the lower waveform of Figure 5.19.

The circuits of Figure 5.21 are very much simplified in order to illustrate the principles involved. Practical integrators and differentiators are normally combined into more complex circuits which include amplifying, clipping and voltage level shifting stages.

The clipping level is reached within the duration of the first field broad pulse in modern integrator circuits. Thus only one field broad pulse is necessary for modern receivers. However, older types require several pulses to reach their clipping level.

5.8 VESTIGIAL SIDEBAND MODULATION

Television services are broadcast in the v.h.f. and u.h.f. bands. The sound and video signals must be modulated on to appropriate r.f. carriers in the transmitter, and the signal reaching the receiver antenna normally goes through several stages of amplification and frequency conversion before being converted back into audio and video frequency signals. The video signal extends from d.c. to between 3 and 10 MHz, depending on the system, (see Table A.1 in the Appendix) and the audio signal extends from about 20 Hz to about 15 kHz. At the transmitter the video and audio signals are modulated onto separate carriers which may be broadcast from separate antennae or which may be combined in a final output stage and broadcast from a common antenna. The difference in frequency between the two carriers slightly exceeds the video bandwidth. All the major broadcast systems use amplitude modulation (a.m.) for the video signal. Either amplitude or frequency modulation is used for the audio signal.

In many ways the simplest form of amplitude modulation is that used on medium and long wave sound broadcasts. The amplitude of the r.f. carrier is made to follow the fluctuations of the modulating audio signal. If the modulating signal occupies a bandwidth B extending effectively down to zero frequency, the spectrum envelope of the modulated r.f. signal extends over a bandwidth $2B$ and consists of two symmetrical sidebands, the upper and lower sideband, each of width B on either side of the carrier. This spectrum envelope is indicated in Figure 5.22.

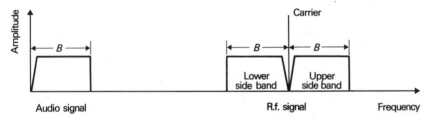

Figure 5.22 Spectrum envelope of an amplitude modulated signal

The advantage of this form of amplitude modulation is that it leads to relatively simple circuits in the transmitter and especially in the receiver, but it is wasteful of bandwidth because either sideband, on its own, contains all the information carried by the modulating signal. It is also wasteful of transmitter power. Most of the power is in the carrier which does not contain any information. The power in each sideband increases with depth of modulation. It reaches a maximum of ⅙ of the total power for 100 per cent modulation. The average modulation depth of broadcast signals is usually much less than 100 per cent, so that the useful power is a considerably smaller fraction than ⅙ in practice.

The greatest saving in both bandwidth and power is obtained by using suppressed carrier single sideband modulation (s.s.b.), as in f.d.m. telephone links, because this only uses a bandwidth B and all the power is in the sideband. Unfortunately, s.s.b. reception had to be ruled out on economic grounds for television because of the complex circuitry needed to restore the carrier in the receiver, and because of the complexity of filter design. Our hearing is insensitive to phase, and the phase characteristics of telephone filters are not critical. The exact shape of video signals must be preserved, and this puts stringent requirements of the phase characteristics needed for s.s.b. television transmission. The form of modulation universally adopted for the video signal in broadcast television systems is **vestigial sideband** (v.s.b.). It involves filtering the r.f. signal so that its spectrum envelope is of the form shown in Figure 5.23, which differs from the normal amplitude modulation spectrum envelope in that most of the lower sideband and most of the carrier have been suppressed. The remaining portion of the lower sideband is called the vestige. It is, of course, possible to attenuate the upper sideband and keep the lower sideband. This is done in the British 405-line system. The 625-line system uses a vestigial lower sideband as in Figure 5.23.

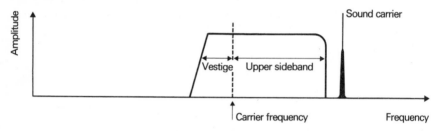

Figure 5.23 **Spectrum** envelope of transmitted television r.f. signals

Figure 5.23 also shows the sound carrier. It is located just beyond the end of the video sideband which is not attenuated, sufficiently far from the band edge so that it can be filtered out together with the audio sidebands before reaching the final stages of the video amplifier. It may be separated out before reaching the video detector, or it may be taken out at a later stage. The demodulation of the audio signal is discussed in Section 5.11.

5.9 QUADRATURE DISTORTION AND THE SHAPE OF THE I.F. AMPLIFIER RESPONSE

Television receivers are normally superhets, most of the amplification before demodulation being carried out in an i.f. (intermediate frequency) amplifier. The i.f. band is, typically, between 33 and 41 MHz and the spectrum envelope of the signal at the i.f. amplifier input is of the form shown in Figure 5.23, but shifted in frequency into the i.f. band. Besides amplifying the signal, the i.f. amplifier acts as a filter whose amplitude/frequency response is shown in Figure 5.24.

The particular shape of Figure 5.24 is chosen to minimise distortion on demodulation, without wasting bandwidth or requiring undue complexity in the receiver circuits, and because it allows envelope detectors to be used which are cheap and simple compared with other possible types. We will now consider envelope demodulation and the way it can introduce distortion when dealing with amplitude modulated signals having portions of their spectrum partially or completely suppressed. This will enable us to specify the optimum shape for the amplitude/frequency characteristic of the i.f. filter.

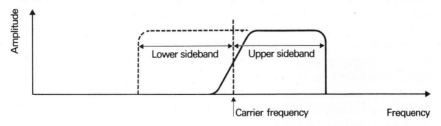

Figure 5.24 Amplitude/frequency response of i.f. filter used for vestigial sideband reception

5.9.1 Envelope demodulation

In order to define terms, we will first consider the envelope demodulation of the double sideband amplitude modulated signal used in long and medium wave sound broadcast systems.

If an r.f. carrier $E_c \cos \omega_c t$ of amplitude E_c and frequency $\omega_c/2\pi$ is amplitude modulated with a sinusoidal signal $E_m \cos \omega_m t$, the resulting r.f. signal, which is shown in Figure 5.25, can be written

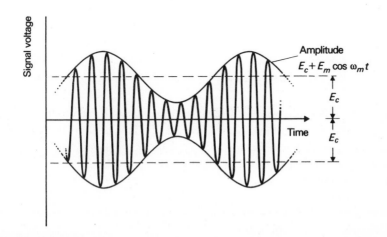

Figure 5.25 Radio frequency carrier with sinusoidal amplitude modulation

181

$$e_{rf} = (E_c + E_m \cos \omega_m t) \cos \omega_c t \qquad\qquad\qquad (5.5a)$$

or equivalently

$$\begin{aligned} e_{rf} = {} & E_c \cos \omega_c t + \tfrac{1}{2} E_m \cos (\omega_c + \omega_m)t \\ & + \tfrac{1}{2} E_m \cos (\omega_c - \omega_m)t \end{aligned} \qquad\qquad (5.5b)$$

where the second and third terms represent the upper and lower sidebands respectively. Throughout this section it is assumed that $E_m/E_c = m \leqslant 1$, where m is the modulation index. This means that the depth of modulation is never more than 100 per cent, which is generally true for television systems.

Modulating signals used in telecommunication do not consist of single sinusoids, but it is convenient, provided the system is approximately linear, to think of the Fourier components which make up the signals and concentrate on one component because this simplifies the analysis.

As shown in Figure 5.26, the carrier and the two sidebands can conveniently be represented by a phasor diagram (Reference 3, Section 3.2; Reference 4, Section 5.2). The phasor representing the carrier appears as a fixed line of length E_c, and the sidebands appear as two phasors of length $\tfrac{1}{2}E_m$ counter-rotating with angular frequency ω_m. The upper sideband phasor, which has a continually increasing phase with respect to the reference phase of the carrier, rotates in the positive direction, which is conventionally taken as anticlockwise.

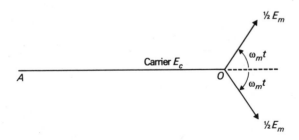

Figure 5.26 Phasor diagram for amplitude modulated r.f. carrier

Envelope demodulation consists in obtaining a signal corresponding to the envelope of the waveform shown in Figure 5.25. This is the vector sum of the three phasors shown in Figure 5.26. It can be deduced, very simply, by resolving each of the two sideband phasors into two components $\tfrac{1}{2}E_m \cos \omega_m t$, parallel to the carrier E_c, and $\tfrac{1}{2}E_m \sin \omega_m t$ normal to it, as in Figure 5.27(a). The normal components are equal and opposite and therefore cancel. The parallel components add to give $E_m \cos \omega_m t$ which, combined with E_c, gives $E_c + E_m \cos \omega_m t$. This is the envelope of the signal shown in Figure 5.25 and is represented in Figure 5.27(b) by the variable phasor AB, with A fixed and B moving to and fro with frequency $\omega_m/2\pi$ between extreme values P and Q corresponding to $E_c + E_m$ and $E_c - E_m$. This is just how $E_c + E_m \cos \omega_m t$, that is the amplitude of e_{rf} in Equation (5.5a), varies with time.

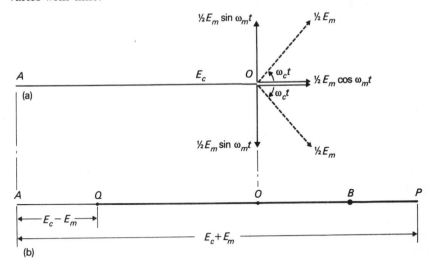

Figure 5.27 (a) The phasors of Figure 5.26 resolved into two pairs or perpendicular phasors. (b) The tip, B, of the resultant phasor moves along line PQ

This result is particularly simple and can be obtained directly by looking at Figure 5.25. The same phasor technique provides a simple way of considering some less obvious aspects of single and vestigial sideband modulation.

5.9.2 Single sideband plus carrier: quadrature distortion

Both the upper and lower sidebands of the signal in Equation (5.5) contain all the modulation information. It should therefore be possible to retrieve this information, even if one sideband were not transmitted in order to economise on bandwidth.

Instead of equation (5.5) we can consider an r.f. signal of the form

$$e_{rf} = E_c \cos \omega_l t + E_m \cos (\omega_c + \omega_m)t \tag{5.6}$$

Compared with the signal of Equation (5.5) the lower sideband has been filtered out and the amplitude of the upper sideband has been doubled. The process of going from double sideband (Equation (5.5)) to single sideband plus carrier (Equation (5.6)) can be thought of as filtering out one sideband and half the carrier, then amplifying the remaining signal by a factor of 2. The reason for this doubling is explained below.

The phasor diagram for this signal is shown in Figure 5.28. It consists of a fixed phasor AO of amplitude E_c (corresponding to the carrier) and a single phasor of amplitude E_m (corresponding to the upper sideband) rotating with angular velocity ω_m about one end of the carrier. At any instant in time the envelope voltage is equal to the amplitude of the resultant phasor AB.

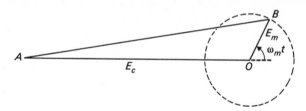

Figure 5.28 Phasor diagram for single sideband plus carrier modulation

The locus of AB is no longer parallel to the carrier; its amplitude can no longer be represented simply by $E_c +$ (a sinusoidal term with frequency $\omega_m/2\pi$). In other words, envelope demodulation no longer provides a signal whose value at any instant is the sum of the modulating signal and a d.c. term equal to the carrier amplitude, that is $E_c + E_m \cos \omega_m t$ as with the previous case (Equation (5.5a)). However, the envelope can be written as

$$AB = E_c + E_m \cos \omega_m t + \Delta \tag{5.7}$$

where Δ is an extra term representing the distortion due to demodulation. The reason for doubling the upper sideband amplitude is that without this the second term in Equation (5.7) would have been $\frac{1}{2}E_m \cos \omega_m t$.

Δ is called **quadrature distortion.** The reason why this term is used can be seen from Figure 5.29(a) in which the modulating phasor OB of Figure 5.28 has been resolved into four rotating vectors, each of magnitude $\frac{1}{2}E_m$: OD_1, and OC_2 which combine to form the original phasor OB, together with OD_2

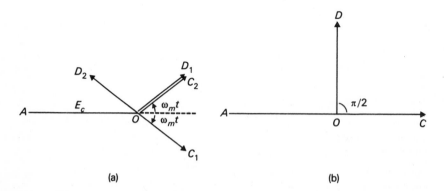

(a)   (b)

Figure 5.29 (a) Phasor diagram equivalent to that of Figure 5.28. (b) Phasor OD represents the quadrature distortion

and OC_1, which are equal and opposite and therefore cancel. Alternatively, these four vectors can be grouped into two counter-rotating pairs. The first pair OC_1 and OC_2 is symmetrically placed about the carrier axis. It is just like the double sideband signal of Figure 5.27. It can therefore be combined to give the phasor OC of Figure 5.29(b) which does not introduce any distortion on demodulation. The second pair OD_1 and OD_2 can be combined into a phasor OD which is in quadrature with OC and is responsible for all the distortion.

The quadrature distortion term Δ can be expressed approximately by

$$\Delta \simeq E_c \, \tfrac{1}{4} \, m^2 \, (1 - \cos \omega_m t) \tag{5.8}$$

where $m = E_m / E_c$ is the modulation index.*

The significant thing about this expression for the quadrature distortion term, Δ, is that it is proportional to m^2, that is it falls off rapidly as the modulation index decreases. Studies of typical spectra of television video signals show that their amplitudes decrease with increasing frequency, so that when a video signal is modulated onto a carrier, the modulation index, m, of sideband components away from the carrier is much smaller than those close to it. It follows from this that the quadrature distortion resulting from the use of single sideband modulation may be acceptable for the high frequency components of the video spectrum, but not for the low frequency ones. The effects of quadrature distortion can therefore be minimised by using vestigial sideband demodulation in which one sideband is entirely suppressed well away from the carrier but is only partially suppressed close in to the carrier, as shown in Figure 5.23.

5.9.3 Vestigial sideband demodulation

Vestigial sideband demodulation is chosen because it allows a relatively simple envelope detector to be used while avoiding the effects of quadrature distortion. As explained in Section 5.9, the shaping of the signal spectrum is

---

*Equation (5.8) can be obtained as follows: from Figure 5.28 the modulation envelope AB is given by
$AB = (OB^2 + OA^2 - 2(OB \times AO) \cos A\widehat{O}B)^{1/2} = E_c (1 + m^2 + 2m \cos \omega_m t)^{1/2}$
This can be expressed approximately by the first three terms of a Fourier series with coefficients a, b and c.
$AB = E_c (1 + m^2 + 2m \cos \omega_m t)^{1/2} = a + b \cos \omega_m t + c \cos 2\omega_m t$
a, b and c can be evaluated by substituting three convenient values of $\omega_m t$ (e.g. 0, $\pi/2$ and π) in turn in the above equation.
If this is done, the expressions for a and c are found to contain the term $(1 + m^2)^{1/2}$ which can be replaced by the truncated binomial expansion $1 + \tfrac{1}{2}m^2$, because we are interested in cases when m is small compared with one. The expression we obtain in this way for AB is
$AB \simeq E_c (1 + \tfrac{1}{4}m^2 + m \cos \omega_m t - \tfrac{1}{4} m^2 \cos 2 \omega_m t)$
Comparing this with Equation (5.7), and remembering that $E_m = mE$, we obtain Equation (5.8) for Δ.

carried out in the i.f. amplifier. The required filter characteristic, that is the frequency response of this amplifier for vestigial sideband demodulation will now be discussed.

We will make the simplifying assumption that the filter has minimum attenuation of 0 dB. In practice the i.f. stages of a receiver consist of amplifiers, as well as filters, and minimum attenuation corresponds to a gain of a few tens of decibels.

The ideal required frequency response is shown in Figure 5.30. It consists of three regions.

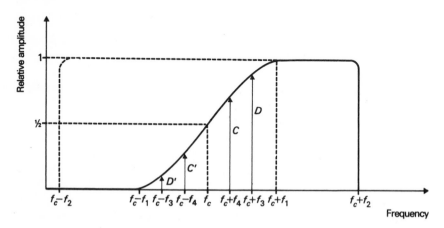

Figure 5.30 Amplitude/frequency response of vestigial sideband i.f. filter. This is an idealised amplitude response. The upper cut-off region of a practical filter is much less sharp so as to prevent ringing and simplify the circuit.

1. $f_c + f_1$ to $f_c + f_2$, where f_c is the carrier frequency. This is the region of minimum attenuation. It consists of most of the upper sideband which, by our simplifying assumption, is therefore not attenuated. The highest video modulating frequency is f_2.

2. $f_c - f_1$ to $f_c - f_2$. This is the region of the lower sideband which is completely attenuated.

3. An intermediate region of partial attenuation between $f_c - f_1$ and $f_c + f_1$ called the **Nyquist flank.** It is important that the response in this region should be approximately antisymmetrical about f_c. To see what this means, consider how a modulating component of frequency f and amplitude E_m is attenuated when passed through the filter. If f is greater than f_1, the corresponding upper side frequency is greater than $f_c + f_1$, this component is not attenuated and its amplitude at the output of the filter is still equal to E_m.

If the modulating frequency is f_3, the upper side frequency is $f_c + f_3$, as shown in Figure 5.30, the component is attenuated and its amplitude at the

186

output of the filter is DE_m , say. The corresponding lower side frequency $f_c - f_3$ is more attenuated, giving an amplitude $D'E_m$. In order to be antisymmetrical, the required filter characteristic must ensure that $(D + D')E_m = E_m$ that is $D + D' = 1$ for all frequencies inside the sloping region. A lower modulating frequency than f_3, f_4 say, would give a value $f_c + f_4$ closer to the carrier with amplitude CE_m which, combined with the corresponding amplitude $C'E_m$, would also add up to E_m . Thus, in the limiting case of zero video frequency, we see that the attenuation at f_c must be such as to halve the signal voltage. That is to say that the filter must have an attenuation of 6 dB at the carrier frequency.

The reason for the antisymmetrical response can be seen by considering phasor diagrams for the various pairs of side frequencies indicated in Figure 5.30.

Consider the pair close to the carrier with frequencies $f_0 + f_4$ and $f_0 - f_4$. After passing through the filter, they are presented to the demodulator with amplitudes CE_m and $C'E_m$. If C were equal to C' we would get the double sideband situation of Figure 5.27 with no distortion. However, $C \neq C'$ and what we get is shown in Figure 5.31(a). The phasor corresponding to the carrier is $\frac{1}{2}E_c$, where E_c is the carrier amplitude before filtering. The resultant of the two sideband components is no longer parallel to the carrier. The locus of the point B is now a narrow ellipse, but the peak to peak variation of the envelope phasor AB takes it from P to Q, corresponding to an amplitude $(C + C')E_m = E_m$ for the demodulated signal. The amplitude is therefore correct, though the waveform is no longer sinusoidal, due to quadrature distortion corresponding to the finite width of the ellipse. The amplitude of the quadrature term is $OR = (C - C')E_m$ which is small near the carrier, that is to say for low video frequencies where the modulation index is greatest.

Figure 5.31(b) shows the situation for frequency f_3 which is further from the carrier, but still on the Nyquist flank. The peak value of the demodulated signal is still E_m, but the quadrature distortion has increased.

For frequencies greater than $f_c + f_1$ we have single sideband demodulation. The resulting amplitude is still E_m. The effect of quadrature distortion is greatest here for a given E_m. However, the value of E_m relative to the carrier amplitude, and hence the modulation index m, decreases with increasing video frequency, so that the effect of quadrature distortion is small, as discussed in Section 5.9.2. The phasor diagram no longer forms an ellipse, but a circle with a much smaller radius compared with the size of the ellipses. This is indicated in Figure 5.31(c).

Thus the shape of the Nyquist flank ensures that the amplitudes of the various Fourier components of the demodulated signal are directly proportional to the corresponding amplitudes in the video signal over the complete video bandwidth. It also ensures that the quadrature distortion is kept within acceptable limits throughout the band.

187

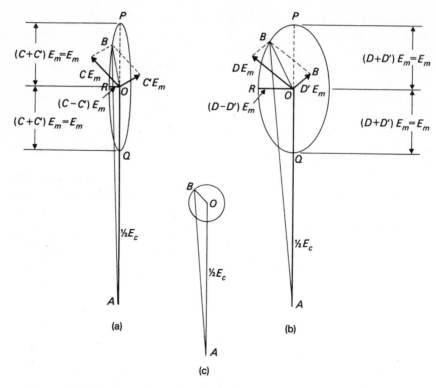

Figure 5.31 Phasors and envelope locus for vestigial sideband demodulation. (a) For a pair of side frequencies on the Nyquist flank and close to the carrier. (b) For a pair of side frequencies also on the Nyquist flank, but further from the carrier. (c) For side frequencies in the passband of the filter, but not on the Nyquist flank

5.10 THE TRANSMITTED SIGNAL

Channel transmission characteristics for the British 625-line system are shown in Figure 5.32. Similar information for other systems can be derived from Table A.1 in the Appendix. The vestigial sideband, which is obtained by means of a filter in the transmitter, extends beyond the band used in the receiver. Shaping to the Nyquist flank is carried out by a filter in the i.f. stages of the receiver. The effect of the i.f. filter is shown by the dashed lines in Figure 5.32. The origin for the frequency scale in the figure is the vision carrier frequency. The signals are transmitted in the u.h.f. band (the 405-line v.h.f. transmissions are due to be converted to 625 lines in the 1980s, and will not be considered here). Ultra high frequency television broadcasts are in band IV (470-610 MHz) and band V (610-940 Mhz). These bands are divided into 8 MHz sections called **channels,** each of which carries one broadcast which, as shown in Figure 5.32, takes up 7·75 MHz, with the upper 0·25 MHz of each channel providing a guard band between adjacent channels. The channels are numbered in order of increasing frequency. The lowest channel in band IV is number 21 (vision carrier frequency $f_c =$ 471·25 MHz). The highest is number 34 ($f_c = $ 575·25 MHz). Channels 35 to 38 have not been assigned in the United Kingdom. Band V starts at channel 39 ($f_c = $ 615·25 MHz) and goes up to channel 68.

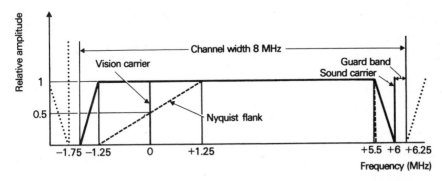

Figure 5.32 Channel characteristics of r.f. transmission channel for the British 625-line system

Figure 5.32 shows that:

(a) The sound carrier is outside the vision band and separated from the vision carrier by 6 MHz. This separation is maintained precisely in the transmitter.

(b) The effective video bandwidth is 5·5 MHz, the upper cut-off of the i.f. filter.

(c) The response of the transmitter filter is flat down to 1·25 MHz below the carrier frequency, and then drops to a negligible value over the next 0·5 MHz.

(d) The Nyquist flank, shown dashed in the figure, extends to 1·25 MHz either side of the vision carrier.

Figure 5.32 shows the response of filters in the transmitter and the receiver. These filters limit the bandwidth of the signal spectrum, but they do not determine its detailed shape.

The spectrum of the signal depends on the type of picture being transmitted, but the video signal has a marked periodic structure because of the sync pulses which repeat regularly at the 15·625 kHz line frequency and at the 25 Hz picture frequency. As a result of this, the spectrum tends to be bunched around harmonics of 15·625 kHz, with a fine structure of lines spaced by 25 Hz.

The spectrum extends over 5·5 MHz, but the majority of the energy is concentrated at the low frequency end for most transmitted pictures. The fact that the monochrome video spectrum does not contain much energy at the high frequency end of the band proves to be an advantage in colour television, as will be explained in Section 6.4.

5.11 RECEIVER STRUCTURE

The basic elements of a receiver are described in this section. The British 625-line system will be chosen to provide a specific example, including numbers. Other systems lead to different numbers but these can be deduced from the data in Table A.1 of the Appendix once the functions of the elements of one type of receiver are understood. The receiver elements are shown in Figure 5.33. They will be dealt with in turn.

The antenna

The country is divided into reception areas with four 8 MHz channels allocated to each area. The channels are chosen so as to minimise interference between neighbouring areas. If channels adjacent in frequency are numbered consecutively and if the lowest channel number in one area is n; then the highest is at most $n + 10$. Hence the four allocated to a particular area are contained in a bandwidth not exceeding 88 MHz. Receiver antennae are designed for optimum sensitivity over this bandwidth. The operating frequency of an antenna depends on its dimensions, so that different areas require aerials of different sizes, but in each area only one is needed in order to receive all four allocated u.h.f. channels.

The u.h.f. tuner

The tuner is made up of an r.f. amplifier and a frequency changer. The frequency changer is usually a single transistor stage acting as local oscillator and mixer. The outputs of the oscillator and the r.f. amplifier are combined in the mixer. These two elements are tuned to select one of the four available channels. Imagine that channel 50, with a vision carrier frequency of 703·25 MHz and a sound carrier frequency 6 MHz higher, that is 709·25 MHz, is selected. The r.f. amplifier is tuned for this channel and the local oscillator is tuned, typically, to be 39·5 MHz above the vision carrier, that is to 742·75 MHz, so that the mixer output contains a difference frequency of 39·5 MHz, corresponding to the vision carrier. This is the intermediate frequency corresponding to a d.c. video signal. The video band is 5·5 MHz wide. A video component of 5·5 MHz produces an r.f. signal of 708·75 MHz and an i.f. signal of 742·75 − 708·75 = 34 MHz, similarly the sound carrier produces an i.f. signal of 33·5 MHz.

190

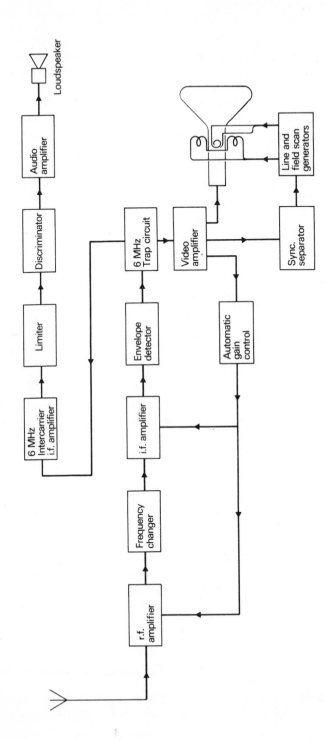

Figure 5.33 Block diagram of a monochrome television receiver

The vision i.f. amplifier

The frequency response of this amplifier is shaped to provide the required Nyquist flank for the modulated vision signal. It also contains a number of trap circuits. These are tuned circuits which sharply attenuate signals over a narrow band. For instance, if we assume again that channel 50 has been selected, the vision carrier of adjacent channel number 51 is at $703 \cdot 25 + 8 = 711 \cdot 25$ MHz. Combined with the local oscillator signal it produces an i.f. signal of $742 \cdot 75 - 711 \cdot 25 = 31 \cdot 50$ MHz, which may be strong enough to cause interference, unless attenuated by a trap circuit tuned to this frequency. Similarly a trap circuit is needed to attenuate the signal due to the sound carrier in adjacent channel 49 which produces an i.f. signal of $41 \cdot 5$ MHz, because the frequency of the sound carrier in channel 49 is $703 \cdot 25 - 8 + 6 = 701 \cdot 25$ MHz. The vision i.f. amplifier also contains a trap to attenuate the sound carrier in channel 50 in order to reduce it to a suitable level for inter-carrier mixing, as described in the next subsection.

The vision detector

The vestigial sideband amplitude modulated vision signal, having been passed through a filter with the appropriate Nyquist flank, can be demodulated using a simple envelope detector. This detector is a non-linear device and it can therefore act as a mixer, producing outputs at frequencies which are the sum and differences of the frequencies of its input signals.

The sound signal is frequency modulated onto a carrier spaced 6 MHz from the vision carrier. The modulated sound carrier and the vision carrier will therefore combine to produce a signal which is frequency modulated onto 6 MHz. This is known as the **inter-carrier sound** signal. It provides an i.f. signal for the sound receiver section.

The sound receiver section

The sound receiver is essentially part of a conventional f.m. sound receiver with a 6 MHz i.f. This i.f. signal is taken from the vision detector via a 6 MHz filter which also acts as a trap circuit to keep the inter-carrier sound signal out of the video circuits.

If an f.m. demodulator tuned to the $33 \cdot 5$ MHz sound i.f., produced by a u.h.f. tuner, were used, any drift in the local oscillator of the tuner would result in the sound intermediate frequency drifting away from the $33 \cdot 5$ MHz centre frequency of the discriminator and this would cause distortion of the sound.

This problem does not arise with inter-carrier working because the 6 MHz frequency difference between the sound and vision carriers is held under tight control in the transmitter and is not affected by local oscillator drift in the receiver.

Video amplitude modulation is removed from the 6 MHz i.f. signal by means of amplification and limiting before it reaches the f.m. sound discriminator. This is not possible in systems such as the British 405 line, which have a.m. sound. They use an i.f. amplifier fed directly from the tuner. The effect of local oscillator drift is reduced by using an i.f. bandwidth of about 200 kHz so as to overlap the 10 to 15 kHz sound bandwidth by a sufficiently wide margin.

The video amplifier

This amplifies the combined video signal in order to bring it to the required level to drive the modulating electrodes of the cathode ray tube and control the beam current, and hence the spot luminance. The tube electrodes are biassed so that the black level corresponds to minimum luminance, the beam therefore being blanked during the sync pulses. Maximum white level corresponds to a video amplifier output of about 50 V above black level. Outputs from the video amplifier are also used to provide synchronisation signals for the scan generators and to drive the automatic gain control (a.g.c.) circuit. These are shown in Figure 5.33. The scan generators were discussed in Section 5.7.1.

The automatic gain control (a.g.c.)

The strength of the received signal may vary due to reflections from passing aircraft and other changes in the radio propagation path. It may also change from one channel to the next because of differences in propagation, transmitter power or antenna sensitivity. The purpose of the automatic gain control is to compensate for these differences so that there is no need to adjust receiver gain when changing channel, or when propagation conditions alter. The a.g.c. circuit operates by extracting, from the composite video signal, a control voltage which depends on the general signal level, independently of specific picture detail. The sync pulse amplitude (in the case of negative modulation) or the back porch level can be used for this purpose. This control voltage is used to modify the operating conditions of the r.f. and i.f. amplifiers so as to alter their gain in such a way as to maintain the same overall video level independently of average signal strength at the antenna.

Only the principal elements of a monochrome receiver have been mentioned in this brief description. Many extra features which can be included in a receiver to give improved performance have been left out. They are discussed in textbooks on television such as References 56 to 58.

5.12 GAMMA CORRECTION AND SYSTEM GAMMA

The subject of gamma correction will be briefly mentioned before moving on to colour television. Gamma correction is used to compensate for the fact that cathode ray tubes and television cameras are inherently non-linear transducers.

193

The luminance, L, of the spot on the screen of a cathode ray tube is given approximately by

$$L = K (E_{gc} - V_c)^{\gamma}$$

where K is a constant, E_{gc} is the voltage applied between the grid and cathode of the tube and V_c is the cut-off voltage, that is the value of the grid-cathode voltage for which the luminance falls approximately to zero. The exponent γ, is the **tube gamma.** It has, typically, a value between 2 and 3. The tube is normally given a d.c. grid-cathode bias equal to V_c, so that $E_{gc} = E + V_c$, say, where the luminance (video) signal applied to the cathode is $-E$. The spot luminance is then proportional to E^{γ}. The relation between the video signal and spot luminance is non-linear because $\gamma \neq 1$.

Television cameras are also non-linear. Their output voltage E_0 is approximately related to the scene luminance L_s by

$$E_0 = K_s (L_s)^{\gamma_s}$$

where K_s is a constant and γ_s is the **camera gamma** which varies between $0 \cdot 3$ and 1 depending on the type of camera tube used.

If the video voltage E in the receiver is proportional to the camera output voltage E_0, the tube luminance L is proportional to

$$[K(K_s L_s)^{\gamma_s}]^{\gamma}$$

It is therefore proportional to $L_s^{\gamma_s \gamma}$

If $\gamma_s \gamma = 1$, the luminance of the displayed picture is proportional to the scene luminance. The exponent $\gamma_s \gamma$ is known as the **system gamma** and it provides a measure of system linearity in so far as luminance is concerned.

A system gamma of between 1 and $1 \cdot 3$ is normally used. User tests have established that $1 \cdot 2$ is generally considered to be the most satisfactory value. The required value is obtained by passing the video signal through suitable non-linear circuits known as **gamma correction** circuits. These could be located in the receiver or the picture source but, since there are many receivers to each picture source, it is more economical to carry out this correction at the picture source.

A further reason for applying the gamma correction at the picture source is that it involves compressing the signal so that high signal levels are more attenuated than low ones. This effectively enhances low level signals and gives a better signal-to-noise ratio than would be obtained with gamma correction at the receiver.

A gamma correction circuit is normally inserted at the camera output. This makes the video signal proportional to $(L_s)^{\gamma_s \gamma_c}$ where the exponent γ_c is due to the correction circuit. The displayed luminance is then proportional to

$(L_s)^{\gamma_s \gamma_c}$ and the system gamma is $\gamma \gamma_s \gamma_c$ or $\gamma \gamma_t$ where $\gamma_t = \gamma_s \gamma_c$ is called the **transmitted gamma.**

Thus with a system gamma of 1·2 and a typical value of 2·8 for γ_s, the gamma correction circuits would have to be adjusted to give a transmitted gamma of 2·8/1·2 $\simeq$ 2·3. The controls of gamma correction circuits are normally calibrated in terms of the transmitted gamma, which gives a measure of the degree of non-linearity between the scene luminance and the magnitude of the video signal.

The choice of system gamma affects the contrast of a black and white picture. As will be seen in Section 6.4.3 it has more complex effects on colour pictures.

Chapter 6

Colour television systems

6.1 THE COLOUR SYSTEMS

There are three colour television broadcast systems in public use known as NTSC, PAL and SECAM. Every national colour television service uses one of these three, though the same system may be implemented slightly differently in different countries.

The letters NTSC stand for the National Television Systems Committee of the USA. This committee coordinated research and development in the American electronics industry. Its achievements are very impressive. Starting from the basic principles of a complex system, which had yet to be implemented in detail, it took only slightly over two years to design and produce prototype equipment and to carry out user tests. The Federal Communications Commission of the USA adopted the NTSC system in December 1953.

The NTSC system is used with 525 picture lines and a 60 Hz field frequency in the USA, Canada, Mexico and Japan.

Most other countries use PAL (Phase Alternation by Line) or SECAM (Séquentiel Couleur à Mémoire) with 625 lines. SECAM was originally developed in France in 1959 and PAL in Germany in 1962. Both were elaborated and modified over subsequent years. They were designed to overcome some defects of NTSC, particularly the effect of differential phase distortion which is discussed in Section 6.9.1. Both were candidates for adoption as the official European colour system, but conflicting views and interests could not be reconciled, with the result that the first European services were started in the United Kingdom and in Germany in 1967 using PAL, and were followed by the French and Russian systems using SECAM.

The purpose of this chapter is to describe how the principal features of the colour systems were chosen. The PAL and SECAM systems evolved from the NTSC system and most of the relevant ideas were elaborated during the development of NTSC. The bulk of this chapter is therefore concerned with the NTSC system. The significant differences between NTSC and the other two systems are described in Sections 6.10 and 6.11.

The design of colour television systems relies heavily on several aspects of visual perception. The signals used are directly related to colorimetric quantities which represent the properties of colour sources as we perceive them. The relevant basic ideas of colorimetry are described in the next section.

6.2 COLORIMETRY AND THE SPECIFICATION OF COLOUR

Any narrow band (monochromatic) light source within the visual range (which was indicated in Figure 5.1 of Chapter 5) is perceived as having a colour. The colours produced by narrow band sources are known as **spectral colours** and are given the names shown in Figure 5.1. **Non-spectral colours** are produced by sources which simultaneously radiate energy from separate regions of the visible spectrum. For instance, a combination of red and blue lights is perceived as a single non-spectral colour, called magenta, a kind of purple. In fact any light source, whatever its spectrum, is perceived as being one colour. However, what is perceived as a particular colour does not have a unique spectrum. Indeed, many different combinations of colours, spectral and non-spectral, can be combined to produce the same perceived colour. In particular, suitable combinations of red, green and blue light sources can be used to create the same visual effect as most of the colour sources which might go to make up a television scene. Red, green and blue are known as **primary colours** and they are used to build up the full range of colours perceived when looking at a colour television picture.

The use of three primary colours is basic to colour television. This is apparent from Figure 6.1 which shows a common arrangement for a television system. Light from the televised scene is separated into three

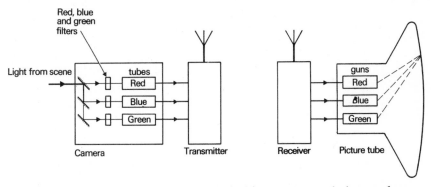

Figure 6.1 Use of primary colours in television cameras and picture tubes

beams, each of which passes through a filter before reaching one of three camera tubes. The first filter transmits only red, the second only green and the third only blue light. The outputs of the three tubes are suitably modulated onto an r.f. carrier which is amplified and broadcast. At the receiver, three signals proportional to the three camera tube outputs are recovered, amplified, and used to control the beam current of three electron guns contained in the envelope of the picture tube, and known as the red, green and blue guns. As described in Section 6.3, the beams from the three guns fall separately on red, green and blue phosphor dots on the screen. At any instant the three beams land on a closely grouped triad of primary colour dots. The beams are scanned, as in monochrome television, except that, instead of a single spot, it is the triads which are used to build up the picture. The three phosphor dots of each triad behave as a combination of primary colour sources which is perceived as having the colour of the corresponding region of the televised scene.

Not every perceivable colour can be produced by combining three primaries. This is discussed in Section 6.2.2. The choice of primaries is not unique. Sets with different shades of red, green and blue can be chosen. Each set allows a slightly different range of colours to be built up. The choice of primaries in television systems depends on the availability of phosphor materials able to produce the required range of luminance for the three colours when bombarded by electron beams having currents and accelerating voltages suitable for domestic receivers.

6.2.1 Colour matching and tristimulus values

In order to be able to see how colour information in a colour television system can be specified in terms of three primary colours, it is useful to consider the colorimeter experiment shown, in principle, in Figure 6.2. A colour source C, which is to be simulated by the light from three primaries, is used to illuminate screen S_1. Screen S_2 is illuminated by the combined light of three primary red, green and blue sources R, G and B. The two screens are viewed simultaneously, in such a way that the two illuminated spots are seen side by side, and the amounts of light from the three primary sources which reach screen S_2 are varied until the illuminated spots on the two screens appear identical. This colour matching procedure is found to have good repeatability and there is sufficient agreement between experimenters to enable standards, representing average results, to be set up. The three primaries are often called **reference stimuli.**

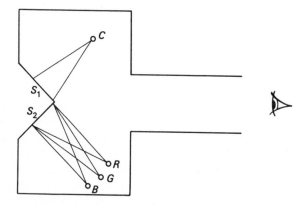

Figure 6.2 Colour matching in a colorimeter

The characteristics, as we perceive them, of colour source C can be specified by the settings of the intensity controls for the three primary sources. These settings can be expressed directly in lumens, or in terms of the luminance they produce on screen S_2. However, it is more convenient to use what are known as **tristimulus values** because, as explained in the next two subsections, this enables one to specify the colour of a source in a standard way, and also because television cameras are normally set up to give outputs proportional to the tristimulus values for the scenes being televised.

The scales of a colorimeter are set up to read in tristimulus values by first setting the luminances of the spots formed by the three primary sources on screen S_2 of the colorimeter to give a match to a standard source of white light, which is called the **reference white**; and then by adjusting the scale factors on the intensity controls of the three sources so that all three give the same reading for reference white. The reference white source is usually an incandescent tungsten filament or a fluorescent lamp operated at a specified current and fitted with filters to give the required spectral distribution.

Once the scales have been set up, other colours can be matched. If a colour C, say, is matched when the red, green and blue scales are set to read R, G and B respectively, then R, G and B are said to be the tristimulus values of colour C.

The luminance of a colour source, which is a measure of its *perceived* brightness, is a quite definite quantity, unambiguously defined in terms of internationally agreed units (candela per square metre). Compared with luminance, the tristimulus values of a colour are highly arbitrary. The three scales from which these values are obtained are calibrated to give readings which are *proportional* to the luminances of the primary spots. However, because of the way in which the instrument is set up, the proportionality constants for the three scales are in general different from each other. They depend on the spectrum and luminance of the reference white and on the particular reference stimuli used in setting up the colorimeter.

You may well be tempted to conclude that measurements involving this degree of arbitrariness cannot be of much use. But this is not the case for the following reasons.

(a) Starting with tristimulus values it is possible to specify a colour match independently of the luminance of the reference white used in setting up the colorimeter. This involves chromaticity coordinates which are discussed in Section 6.2.2.

(b) It is possible to relate the results of colour matching experiments carried out with different reference stimuli and reference white sources. This can be done by means of chromaticity diagrams as explained in Section 6.2.3.

These possibilities arise because it is found experimentally that the results of colour matching experiments for different colours can be combined algebraically. This can be expressed as follows.

If a colour source is matched when the red, green and blue scale of a colorimeter read R_1, G_1 and B_1 respectively, we can write

$$C_1 = R_1 + G_1 + B_1$$

where C_1 is simply the sum of the tristimulus values for the colour. It is the number of tristimulus units produced by the source. This number depends on the choice of reference stimuli and of reference white.

Similarly the match to a second source, obtained using the same reference sources, can be written

$$C_2 = R_2 + G_2 + B_2$$

The experimental result, which is known as **Grassman's law,** can now be expressed by saying that the tristimulus values obtained by matching the combined light sources are found to be $R_1 + R_2$, $G_1 + G_2$ and $B_1 + B_2$, and that the expression

$$C_1 + C_2 = R_1 + R_2 + G_1 + G_2 + B_1 + B_2$$

is found to represent the match of the combined sources.

This result only holds if the colorimeter is set up with the same reference stimuli and reference white in both cases. However, one can match one set of reference stimuli to another by means of colorimeter experiments, and then use the linear relation expressed by Grassman's law in order to write down equations linking the results of colour matching experiments made with different reference sources. It is therefore possible to set up agreed standards to which the results of specific colour matching experiments can be referred. This is discussed further in Section 6.2.2. Grassman's law also holds for colour match equations with both sides expressed in terms of luminance or directly in terms of radiated watts per unit area.

6.2.2 Chromaticity and the colour triangle

It is possible to perceive two light sources as having the same colour but different brightnesses. Thus it is possible for two sources to have the same colour but different luminances. Conversely it is possible for two sources to have the same luminance but different colours. It is useful, in television systems, to be able to separate out the luminance and colour of a source. A colorimeter can be used to obtain the tristimulus values R, G and B, say, of a colour, but these values depend on the luminances of both the colour source and the reference white source. However, the tristimulus values can be used to define three quantities, r, g and b, known as the **chromaticity coordinates** or **chromaticity** of the source, given by

$$r = \frac{R}{R+G+B}, \quad g = \frac{G}{R+G+B}, \quad b = \frac{B}{R+G+B} \qquad (6.1)$$

which are independent of both luminances. This follows from the definitions of r, g and b. If the luminance of the source being matched is changed by a factor, a, say, then, by Grassman's law, R, G and B must change by the same factor and r, g and b remain unchanged. Also, if the luminance of the reference source used in setting up the colorimeter is changed by a factor a, the scale factors for R, G and B must be changed in the same proportion and r, g and b once more remain unchanged.

Hence the chromaticity is independent of the luminance of the colour spot and of the luminance of the reference white spot.

It follows directly from Equation (6.1) that

$$r + g + b = 1 \qquad (6.2)$$

so that only two of the chromaticity coordinates are independent and it is possible to plot chromaticity on a two-dimensional map, such as that of Figure 6.3, which is known as a **colour triangle.** Every point inside the triangle represents a different perceived colour. Its chromaticity coordinates

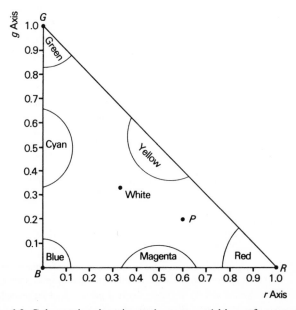

Figure 6.3 Colour triangle using red, green and blue reference stimuli

can be read off using side BR as the r axis and side BG as the g axis. The third coordinate is $b = 1 - r - g$. A triangle is obtained because Equation (6.2) must hold, and because the chromaticity coordinates cannot be negative in the colour matching experiments of Figure 6.2. The three sides therefore correspond to the limiting cases $r = 0$, $g = 0$, and $b = 0$; but $b = 0$ implies that $r + g = 1$, which represents a straight line through points R and G in Figure 6.3.

The point R represents a pure red identical to the red reference stimulus. The point marked white has coordinates, $r = 1/3$, $g = 1/3$, and hence $b = 1/3$. It represents the reference white. The point P has $r = 0{\cdot}6$, $g = 0{\cdot}2$ and hence, $b = 0{\cdot}2$. It can be thought of as a mixture of white and red, that is a diluted form of red, namely pink.

By mixing the three reference stimuli it is only possible to produce perceived colours whose chromaticity coordinates lie between 0 and 1. This means that all perceived colours which can be formed using the three reference stimuli R, G and B must lie inside the triangle R-G-B.

Not all visible colours can be matched by a combination of the three selected primaries. Some can only be matched by transferring one of the primaries to the same side of the colorimeter as the colour being matched, or by using white light (lamp W in Figure 6.4) to dilute the colour. Grassman's law allows us to handle the equations algebraically and the net result is that colours which cannot be matched directly can still be expressed in terms of chromaticity coordinates referred to three primary reference stimuli, but with some coordinates negative, or greater than 1.

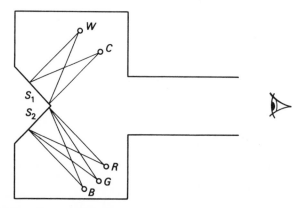

Figure 6.4 Colorimeter using white source W to dilute colour C which cannot be matched directly with the primaries R, G and B.

Thus if a colour source has to be mixed with red light before it can be matched to a mixture of red and green, and if R, G and B tristimulus units are needed to match C tristimulus units of the source, then we can write

$$R + C = B + G$$

that is

$$C = -R + B + G$$

The corresponding chromaticity coordinates are

$$r = \frac{-R}{-R+B+G}, \quad b = \frac{B}{-R+B+G}, \quad g = \frac{G}{-R+B+G}$$

$-R + B + G$ is positive, because C is positive. Hence r is negative, b and g are positive.

For example, $r = -0{\cdot}4$, $g = 0{\cdot}3$, $b = 1{\cdot}1$ corresponds to a colour close to the point labelled 480 in Figure 6.5.

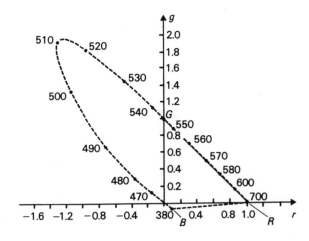

Figure 6.5 Chromaticities of the spectral colours plotted with respect to red and green and blue reference stimuli. The wavelengths of the spectral colours are indicated in nm.

The result of colorimetric matching experiments for all spectral colours is shown in Figure 6.5. The points representing all the spectral colours lie on the dotted horseshoe locus which is labelled with the wavelengths of the spectral colours in nm. All other colours which we are able to perceive are represented by points inside the horseshoe. The line BR joining the ends of the horseshoe, just beneath the r axis, consists of non-spectral colours.

6.2.3 CIE chromaticity diagrams
The chromaticity coordinates of any colour in a diagram, such as that of Figure 6.5, depend on the particular choice of reference stimuli. Standards

have been set by the CIE (Commission Internationale de l'Eclairage). One such set of standards uses fictitious non-physical primary colours X, Y and Z, chosen so that all perceived colours have chromaticity coordinates which are positive and less than 1. The standard primaries are fictitious because they lie outside the horseshoe locus enclosing all physically realisable colours. However, any perceived colour can be matched to these standards by using the process discussed in conjunction with Figure 6.4 with red, green and blue primary sources, and by converting algebraically to CIE coordinates by means of Grassman's law.

The CIE diagram using XYZ primaries is shown in Figure 6.6. The XYZ set was used in laying the foundations of NTSC, the first major colour system, and, although other standard sets have been subsequently introduced, it is a convenient one to use when discussing the choice of system parameters based on properties of colour perception.

Using XYZ primaries we can specify the chromaticity of any colour and, in particular, the coordinates of the primaries used in picture tube phosphors and camera filters. The primaries for the NTSC and PAL systems are shown in Figure 6.6. The CIE x and y coordinates of the red, green and blue

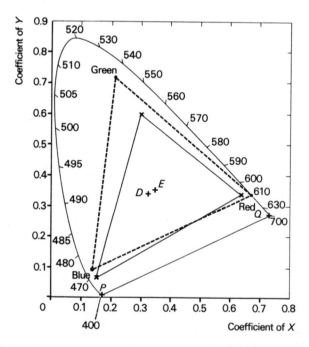

Figure 6.6 CIE chromaticity diagram showing the display primaries for the NTSC (●) and PAL (x) systems. Point E represents the chromaticity of equal energy white and point D represents the chromaticity of standard illuminant $D_{6\,500}$. The chromaticities of the spectral colours are also shown on the diagram

primaries for the NTSC system are red: $(0\cdot67, 0\cdot33)$, green: $(0\cdot21, 0\cdot71)$ and blue: $(0\cdot14, 0\cdot08)$ respectively. The corresponding coordinates for the United Kingdom PAL system are red: $(0\cdot64, 0\cdot33)$, green: $(0\cdot29, 0\cdot60)$ and blue: $(0\cdot15, 0\cdot06)$. The z coordinate is always equal to $1 - x - y$, since we are dealing with chromaticities.

It was mentioned, in Section 6.2.1, that the outputs of cameras are arranged to be proportional to tristimulus values of the scene. This is achieved by using a reference white source. The outputs of the three colour tubes of a camera are separately amplified and the camera is set up by adjusting the gains of these amplifiers so that all three produce equal outputs for a reference white source. These sources are called standard **illuminants.** Illuminant E (point E in Figure 6.6) is the standard used for the CIE diagram. It therefore has CIE chromaticity coordinates $x = y = 1/3$. It is a theoretical standard for which every wavelength of the visible spectrum appears with equal energy. Illuminant $D_{6\,500}$, which is meant to have the quality of normal daylight, is specified as a practical standard for television. Its coordinates are $x = 0\cdot3127$, $y = 0\cdot3290$ (point D in Figure 6.6).

Illuminant $D_{6\,500}$ can be produced in the studio by means of special electric lights having a broad spectrum with a fairly uniform distribution of energy density (about a 2:1 variation over the visible range of wavelengths). The subscript 6500 is used because the spectrum is similar to that of the radiation produced by a perfect black body at a temperature of 6500 K, though a $D_{6\,500}$ source produces somewhat more energy at the short wavelength end of the spectrum.

6.2.4 Tristimulus values and luminance; the luminosity coefficients

The purpose of this section is to relate the tristimulus values of a colour source to its luminance. This is something we need to know because, as will be explained in Section 6.4, video signals are transmitted as a combination of luminance signals and signals carrying the colour information, whereas the cameras are set up in the studio so that the outputs of their three colour tubes are proportional to the tristimulus values of the colour sources in the scene being televised.

A camera and a colorimeter are set up in a similar way using a reference white source. Let us first consider the colorimeter.

Imagine that the reference white source produces a spot luminance of one cd m^{-2} and that, in order to match it, the red, green and blue sources have to be adjusted to produce spots of luminances l, m and n cd m^{-2} respectively. Using Grassman's law we have

$$l + m + n = 1 \tag{6.3}$$

where the quantities on both sides of the equation are in luminance units.

In setting up this colorimeter the red, green and blue scales are adjusted so that they each read one tristimulus unit when a match to the one cd m⁻² reference white spot is obtained. Thus one tristimulus unit of red corresponds to a red spot luminance of l cd m⁻² and, similarly, one tristimulus unit of green or blue corresponds to spot luminances of m or n cd m⁻² respectively.

If, once the colorimeter has been set up, a match to a colour C is obtained with R tristimulus units of red, G of green and B of blue, then the luminance, L_C of the spot produced on screen S_1 of the colorimeter is given by

$$L_C = lR + mG + nB \text{ cd m}^{-2}$$

The quantities l, m and n are known as the **luminosity coefficients.** They represent the luminance of one tristimulus unit of each of the primaries when the colorimeter is set up with a reference white spot of unit luminance.

Now consider a camera which has been set up so that all three of its tubes produce equal output voltages when it is focused on a reference white source. If the tubes behave linearly, the output voltages, E_R, E_G and E_B of the red, green, and blue tubes will be proportional to the tristimulus values of the scene being televised and the voltage

$$E_Y = lE_R + mE_G + nE_B \tag{6.4}$$

obtained by combining the three tube outputs in the proportion of the luminosity coefficients will be proportional to the luminance of the scene. This voltage E_Y is equivalent to the video signal in monochrome television.

The values of l, m and n depend on the reference stimuli used for the system and these are determined by the phosphors chosen for the screens of the receiver picture tubes. The filters used in front of the three camera tubes are chosen to make the tubes sensitive to the appropriate range of frequencies corresponding to the three reference stimuli.

There are slight differences between the chromaticities of the reference stimuli used for the various national television systems but these do not lead to significant differences in the values of the luminosity coefficients which are $l = 0 \cdot 30$, $m = 0 \cdot 59$ and $n = 0 \cdot 11$ for all the main systems.

6.2.5 Chrominance

The output voltages E_R, E_G, E_B of the red, green and blue tubes of a camera are proportional to the tristimulus values of the colour of the scene being televised. The voltage $E_Y = lE_R + mE_G + nE_B$ is proportional to the *luminance* of the colour. A reference white having the same *luminance* would produce equal outputs, E say, from each camera with $lE + mE + nE = E_Y$, the same luminance as before. Using Equation (6.3), it follows that $E = E_Y$.

The chrominance of a colour is defined as the difference between its tristimulus values (represented by E_R, E_G and E_B) and the tristimulus values of a reference white of the same luminance (each represented by $E = E_y$). Hence the chrominance values for the colour being televised are proportional to the signals $E_R - E_Y$, $E_G - E_Y$ and $E_B - E_Y$. Note that, from the definition of chrominance, a grey source has zero chrominance values and that the chrominance of a black and white (monochrome) television picture is zero.

6.2.6 Hue and saturation

As explained in Section 6.2.2, the chromaticity of a colour source is measured independently of its luminance. In terms of visual perception, the properties of a source can be specified by its luminance and its chromaticity. Luminance was discussed in Section 5.1. It is specified by a single number. Chromaticity is specified by two numbers, but, unlike luminance which gives a measure of what we perceive to be the brightness of a source, the significance of these two numbers is not obvious in everyday terms. However, chromaticity coordinates can be related to hue and saturation which come nearer to our subjective impressions of colour.

The meaning of these terms can be seen from Figure 6.7. A point such as P represents a green colour and is said to have a green hue, but so do all other

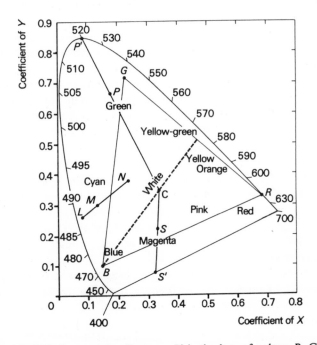

Figure 6.7 CIE chromaticity diagram. P' is the hue of colour P. Colour M can be produced by combining colours L and N in the ratio (luminance of N)/(luminance of L) = LM/NM. The points R, G and B represent the NTSC picture tube primaries

points in the region labelled green. In order to define the colour corresponding to point P more precisely, a line is drawn through P and the point C representing reference white. This line intersects the locus of pure mono-chromatic sources (which lies along the boundary of the range of colours we are able to perceive) at P'. The colour at P' is called the **hue** of colour P. It is a pure green of wavelength 520 mm. This is also called the **dominant wavelength** of colour P. All colours on the line CP' have the same dominant wavelength.

It can be shown, using Grassman's law, that any colour, such as M in Figure 6.7, can be produced by a suitable mixture of colours L and N, where L, M and N lie in the same straight line on the chromaticity diagram, and M lies between L and N. In fact, the required amounts of L and N are proportional to the masses of weights located at L and N which would have their centre of gravity at M.

This property is used to define **saturation.** A colour such as P can be thought of as a mixture of pure colour P' and white light C, where P lies on the straight line joining P' to C. The closer P is to P', the larger the proportion of colour P'. The colour P' is said to have 100 per cent saturation. P is said to be a desaturated colour of hue P'. The degree of saturation of P can be expressed as ($100\ CP/CP'$) per cent. It is sometimes called the **purity** of P.

It is worth remembering that, on a chromaticity diagram, the *hue* of colour P, is determined by the intercept P' of the line CP on the boundary of the horseshoe area. It therefore depends on the *direction* of the line CP. Hence, because the saturation of P is determined by the *length* of the line CP as a fraction of CP', the hue and saturation are roughly equivalent to the coordinates of a polar diagram; that is to the angle and direction of a vector with origin C ending at P. The analogy would be closest if, instead of a horseshoe, the chromaticity of all colours lay inside a circle with centre C.

Not all colours have hues lying on the curved part of the perimeter of the horseshoe area. Some lie on the straight part. The colour S, for instance, has a hue S' which is magenta, and not a spectral colour. We can therefore talk of the hue of S, but not of its dominant wavelength.

A colour triangle formed using the three NTSC primary colours, red, green and blue is drawn in Figure 6.7. A line drawn through blue and white intersects the triangle at yellow. From this we conclude that a colour perceived as white can be produced by combining blue and yellow. Yellow is said to be the **complementary colour** to blue. A complementary colour is one which added in suitable proportions to a primary colour produces a result perceived as white.

6.2.7 The perception of colour change
Consider two light sources of different colours. The extent to which these two colours are perceived to be different is related to the separation of the points

representing them on a chromaticity diagram. The greater the separation, the more different the colours appear. However, the difference we perceive also depends on where the points are in the diagram and on the direction of the change in going from one to the other. We are much more sensitive to changes in the blue region of the chromaticity diagram than in the green. This is illustrated in Figure 6.8. The diagram is divided into oval regions drawn so that the apparent colour change in moving from the central dot in any region to the boundary of that region is the same for all regions. In the NTSC system this variation of visual sensitivity to changes in chromaticity is taken into account when choosing the way in which the camera outputs are combined so as to provide chrominance information which can be modulated onto the r.f. carrier and extracted again at the receiver. Two chrominance signals are used, corresponding to colour changes in the directions of maximum and minimum sensitivity. The two signals are allowed different bandwidths, the signal corresponding to maximum sensitivity having the larger. This point is discussed again in Section 6.6.3.

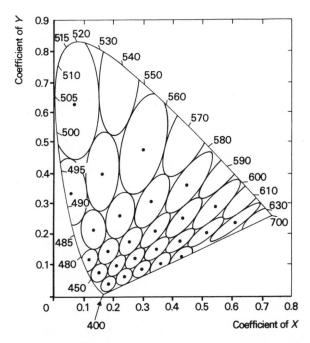

Figure 6.8 The apparent colour change in moving from the central dot to the boundary of any oval region is the same for all regions

6.3 COLOUR DISPLAY DEVICES

Colour television receivers use cathode ray tubes for displaying the picture. Colour tubes are the most expensive and bulky elements of receivers and their correct operation depends on many delicate adjustments which take up a large proportion of servicing time, and hence contribute significantly to servicing costs. Extensive research is being carried out both in order to produce improved tubes, requiring less complex circuitry or simpler adjustments, and in order to develop completely new types of devices which might be cheaper, less bulky or require lower operating voltages than cathode ray tubes. At the time of writing, no clear alternative to the cathode ray tube has yet emerged.

6.3.1 The shadow mask tube

There are several types of colour tube, all of which use three different phosphors, each producing one of the three primary colours when bombarded by an electron beam. The dot triad shadow mask tube, developed by RCA in the United States, is the one currently in most common use.

It will be described briefly in order to illustrate the kind of requirements a colour tube makes on the design and adjustment procedures of a receiver. The description will be limited to meeting this objective which does not include a study of the types of circuits used to correct picture faults, or a detailed description of how receiver adjustments can be carried out in practice. This can be found in References 62 and 64.

The screen of a shadow mask tube is coated with red, blue and green phosphor dots arranged in equilateral triangles, each triangle forming a triad of red, blue and green dots, as shown in Figure 6.9. There are about half a million triads distributed over the screen. Inside the tube (Figure 6.10), and about 1 cm away from the screen, is the shadow mask, a steel sheet about 0·2 mm thick, perforated with as many holes as there are phosphor dot triads on the screen, that is one hole for each triad of three phosphor dots. In the neck of the tube are three electron guns, uniformly placed on a circle about the tube axis. They are the red, green and blue guns. The holes in the shadow mask are located in such a way that, when properly adjusted, the red beam, that is the electron beam from the red gun, can only bombard red phosphor spots. Similarly the green and blue beams can only bombard spots of their colour, as indicated in Figure 6.10. The guns are equally spaced at 120° around the tube axis. The electron beams, after leaving the guns, pass through the magnetic field produced by the scanning coils. This is the deflection field and, if the tube is operating correctly, the beams appear, at the screen, to come from three points symmetrically located at 120° intervals about the axis in a plane at the centre of the deflection field. These points are called the **colour centres.**

It may appear unlikely, at first sight, that colour tubes could be mass produced with such precision that, whatever the deflection, the electron beams would be restricted by the holes in the shadow mask so as to fall only on phosphor spots of the appropriate colour. It is therefore worth mentioning how this is achieved.

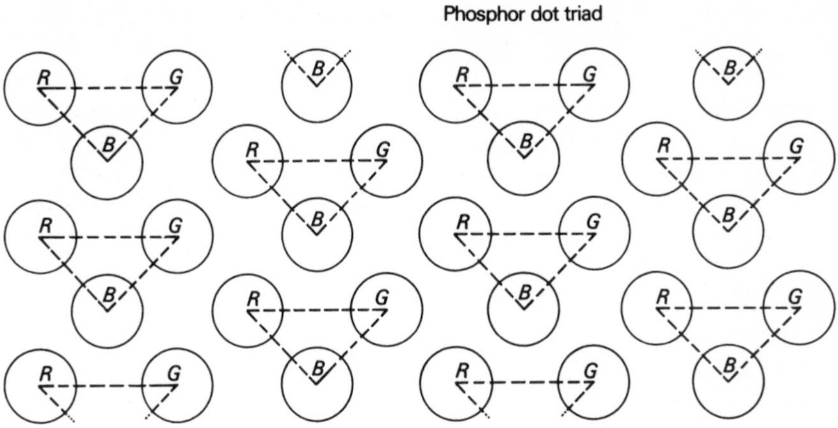

Phosphor dot triad

Figure 6.9 Arrangement of phosphor dot triads in a shadow mask tube. Red, green and blue dots are labelled *R*, *G* and *B*, respectively

During manufacture of the tube, an ultraviolet light source is located at each colour centre in turn and made to illuminate the screen through the shadow mask. The light beam only lands on those regions of the screen which will subsequently be illuminated by the corresponding electron beam. The screen, at this stage, is coated with a photosensitive colour phosphor which, after exposure to light and suitable development, leaves a pattern of phosphor dots on the inner face of the cathode ray tube. The process is carried out three times, once for each colour, the particular shadow mask which will eventually be fitted to that tube being used each time. This ensures that an electron beam deflected at its colour centre can only reach phosphor spots of a single colour. However, this requires that the centres of the deflection fields should coincide with the colour centres. To achieve this condition, **purity control** magnets are mounted on the neck of the tube and adjusted as part of the setting up procedure during the final stages of receiver manufacture, and also, occasionally, during the life of the receiver, in order to compensate for ageing effects.

Two other adjustments, for grey scale tracking and for convergence, which also have to be made during manufacture and servicing will now be described.

Grey scale tracking
The luminance produced by each phosphor dot depends on the current in the electron beam producing it, which, in turn, depends on the voltage, E_{gc} applied between the grid and cathode of the appropriate gun. As will be explained in Section 6.4.2, colour television systems are designed so that monochrome grey scenes produce equal values of E_{gc} at all three guns, when, ideally, the three colours combine so as to produce a grey picture

211

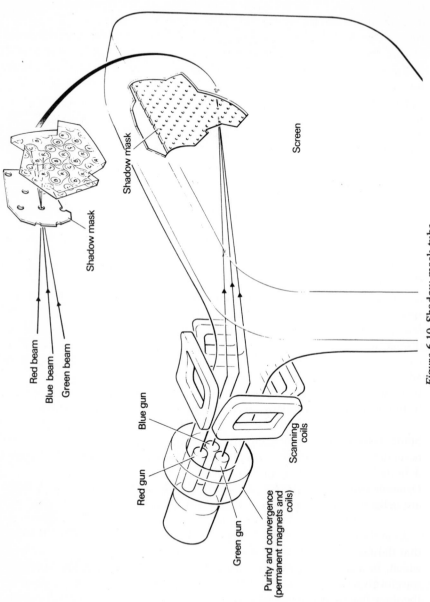

Red beam

Blue beam

Green beam

Shadow mask

Shadow mask

Screen

Red gun

Blue gun

Green gun

Scanning coils

Purity and convergence (permanent magnets and coils)

Figure 6.10 Shadow mask tube

whose luminance depends on the magnitude of E_{gc}. In practice, tinted pictures may be obtained because the phosphor efficiencies (luminance/beam current) differ for the three phosphors, or because a given value of E_{gc} leads to different values of beam current in the three guns. As explained in Section 5.12, the beam current is proportional to $(E_{gc} - V_c)^\gamma$, where V_c is the cut-off voltage corresponding to zero current, and the grid is biased to cut-off (held at a d.c. level of $- V_c$ in the absence of a signal) so that black level corresponds to $E_{gc} = 0$. The current voltage laws of the three guns may vary somewhat both in the value of the cut-off voltage and in the value of γ.

The effect of the cut-off voltage is most significant at low light levels. For instance, if the blue and green guns are just cut-off but the red gun is just above cut-off, a red tint will be produced. The cut-off voltages can be made equal by varying the bias to the first anode of each electron gun. The effect of variations in γ and in phosphor efficiency, which are more noticeable at high luminance levels, can be compensated for by adjusting the drive levels to the grids of the three guns.

There are thus six controls which need to be preset in order to provide what is known as **grey scale tracking** over the full luminance range of monochrome pictures.

Convergence

A correct adjustment of purity ensures that the red beam only falls on red spots, etc., but it does not ensure that, at any instant, the three beams are falling on the three spots of the *same* triad. The beams can be relatively displaced so as to produce three coloured pictures which do not overlap correctly. The process of ensuring that the three rasters coincide is known as adjusting the **convergence.** This is done by applying corrective magnetic fields separately to the three guns, in a region where the guns are shielded from one another by magnetic screens. There are two types of correction:

Static convergence, which is applied, usually by means of adjustable permanent magnets, to the centre of the screen, where the beams are not deflected.

Dynamic convergence, which is applied by electromagnets, when the beams are deflected away from the centre of the screen.

As in the case of the monochrome tube discussed in Section 5.7.1, the fact that the surface of the screen is not spherical causes distortions of the raster which, in a monochrome tube, can be corrected by shaping the scanning waveforms so that they are no longer linear. Unfortunately, in a colour tube, the three beams are deflected about different points and require differently corrected waveforms. Independent line and field corrections have to be applied by means of convergence coils mounted over the neck of the tube in a region where the guns are magnetically shielded from one another. Each gun has two sets of coils, a field and a line coil, so that there are six sets of coils in all. Different waveforms are applied to each set. Each waveform is made up by combining a parabolic and a sawtooth waveform, the proportions of each being determined by two preset controls. Dynamic convergence therefore typically requires twelve controls to be adjusted.

With three sets of static magnets and twelve dynamic preset controls it is not surprising that convergence adjustments are both laborious and skilled, and therefore costly, both in manufacture and maintenance. A number of alternative types of tube with simpler convergence controls have been produced, some of which are described in References 57, 61, 62 and 64.

6.4 THE CHOICE OF VIDEO SIGNALS

This section deals with general considerations leading to the choice of video signals in the three major groups of colour television broadcast systems: NTSC, PAL and SECAM. These considerations are common to all three systems and take us as far as we can go without having to look at the systems separately. The form taken by the video signals carrying the luminance and chrominance signals, including the effects of gamma corrections, will be considered. However, the way in which these signals can be combined into a composite video signal is different in the three systems and will be left to subsequent sections.

6.4.1 Compatibility and reverse compatibility

All broadcast colour television systems have been developed for use in countries which already had monochrome systems. This presented a major constraint on the design of colour systems. Owing to the shortage of broadcast bandwidth they had to operate over the same frequency channels as the existing monochrome systems. This immediately led to two requirements for the new systems.

(a) Existing monochrome receivers should produce satisfactory black and white pictures when receiving a colour transmission. This is known as the **compatibility** requirement.

(b) A standard monochrome transmission should produce a satisfactory black and white picture on a colour receiver. This is known as the **reverse compatibility** requirement.

As we shall see, perfect compatibility is not achieved in any of the current systems, but they all produce compromises which are considered adequate on the basis of user tests.

Probably the simplest to conceive and, at first sight, the most obvious way to transmit a colour signal would be to treat the outputs of the red, green and blue camera tubes as separate monochrome channels with, in effect, three independent transmitters and receivers, the video output of each receiver being applied to one of the three guns in a picture tube of the shadow mask type. This would require three times the bandwidth of a monochrome channel, and even though a monochrome receiver tuned to one of the channels would produce a picture, its luminance would, in general, be completely different from that of the scene. Such a system would not therefore be compatible.

All existing colour broadcast systems operate within the same frequency bands as the corresponding monochrome systems with which they are compatible. They achieve compatibility by separating out the luminance and chrominance information of the colour signals.

The luminance information is converted into a signal which is essentially the same as the luminance signal for a monochrome transmission of the same scene. In particular, it occupies the same bandwidth, which does not seem to leave any room for the chrominance information.

The chrominance information is nevertheless transmitted in the same band. This is made possible because of the following considerations.

(1) Imagine that your are colouring a black and white photographic print. A satisfactory effect can be obtained even when the boundaries of the colours areas are much less sharp than the underlying black and white picture. This is because the eye is less sensitive to colour changes than to changes in contrast, that is changes in luminance from one part of the picture to another. In other words the amount of detail we have to add in the form of chrominance information is much smaller than the luminance detail present in the black and white picture. As we saw in the previous chapter, the bandwidth required for a video signal increases with the amount of picture detail it has to convey. We can therefore conclude that less bandwidth is required for chrominance than for luminance information.

(2) Although a monochrome luminance signal may occupy a bandwidth of, say, 5·5 MHz, it does not occupy the whole of this bandwidth uniformly. As mentioned in Section 5.10, the majority of the spectral energy in the average luminance signal is concentrated at the low frequency end of the video band. A relatively narrow band chrominance signal, having frequency components in the high frequency region of the luminance spectrum, can be combined with the luminance signal so as to have little effect on the luminance of television pictures as perceived by an average viewer. This will be discussed further in Section 6.8.

There is thus room for the chrominance signal in the video spectrum. Furthermore, we can see that reverse compatibility is possible, provided that colour receivers are designed so that grey corresponds to a zero amplitude chrominance signal. If this is the case, a monochrome transmission appears to a colour receiver as one with zero chrominance signal. The receiver therefore produces a grey, that is monochrome, picture of the appropriate luminance.

As for compatibility, this can be achieved provided the extra chrominance signal has no effect on the luminance produced by a monochrome receiver. This is sometimes referred to as the **constant luminance principle,** which can be stated as the requirement that the luminance should be independent of (that is remain constant for) any changes in the chrominance signals and any noise superimposed on these signals.

Experiments have shown that the eye is more sensitive to random fluctuation (noise) in the luminance signals than in the chrominance signals. Thus the constraints set by permissible noise levels in the chrominance signals are less stringent if the noise in these signals is prevented from producing any fluctuations in the luminance.

6.4.2 The colour difference signals and the luminance signal

In this section we will consider the separate forms of the luminance and chrominance signals at video frequency. These forms are essentially the same in the three principal systems, NTSC, PAL and SECAM. The significant differences between the systems arise in the way in which the signals are combined into a composite video signal and modulated onto a carrier at the transmitter, and the way in which they are separated out at the receiver by a variety of demodulation processes.

Figure 6.11 shows how three colour signals can be applied to a receiver picture tube. Consider the red gun. It has a grid-cathode voltage $E_{gc} = E_R + V$, where V is the d.c. bias applied to the grid. As explained in Section

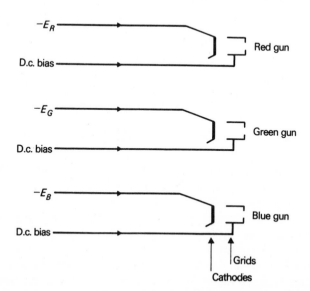

Figure 6.11 Picture tube signals. This is a highly simplified diagram. In particular, preset controls of luminance scale factors and bias voltages have been omitted

5.12, the luminance produced by this grid-cathode voltage, which, in this case, depends on the luminance signal for the red component of the picture, is equal approximately to $K (E_{gc} - V_c)^\gamma$ and, if the d.c. bias voltage V is set equal to V_c, the luminance of this red component of the picture is approximately equal to KE_R. Initially, we will ignore the effects of gamma corrections and assume that $\gamma = 1$, because this simplifies the argument without detracting from the basic ideas. We will return to other values of gamma further on, but at this stage we will assume that the luminances of the red, green and blue components of the displayed picture are proportional to the signals E_R, E_G and E_B respectively, provided the appropriate d.c. bias has been applied to the grids.

Another simplification which we will adopt is to ignore the proportionality factors which apply to various parts of the system. For instance, the signal corresponding to E_R may be of the order of 1 V at the output of the camera, and -50 V at the cathode of the receiver picture tube. The signal corresponding to E_R is amplified or attenuated at various stages through the system but we will ignore this fact and use the same symbol E_R for the signal at the picture tube, at the transmitter, and anywhere else when we deal with it in the system. We can do this provided that the corresponding signals E_G and E_B, and the luminance signal E_Y have been amplified or attenuated by the same amount. We are effectively normalising the signals in order to avoid having to write numerous proportionality constants whose values do not affect the validity of the arguments which will be used.

The complete picture information is contained in the three signals E_R, E_G and E_B. It is these three signals which are produced by the camera and which are applied to the picture tube cathodes; but these are not the signals which are transmitted because, if they were, the system would not be compatible with monochrome television.

The transmitted signals are a **luminance signal** E_Y of the form

$$E_Y = 0 \cdot 30 E_R + 0 \cdot 59 E_G + 0 \cdot 11 E_B \tag{6.5}$$

and a pair of **colour difference signals:** the red colour difference signal $(E_R - E_Y)$ and the blue colour difference signal $(E_B - E_Y)$.

Three signals have to be transmitted because the information they convey corresponds to three independent signals: E_R, E_G and E_B. The luminance and the two colour difference signals are obtained by combining the camera output signals E_R, E_G and E_B in a **matrix,** which is a network of resistors and amplifiers. The luminance and colour difference signals obtained in the receiver after demodulation of the transmitted r.f. signal are combined in another matrix in order to produce tube signals E_R, E_G and E_B.

As explained in Section 6.2.4, E_Y is proportional to the luminance of the scene being televised. Equation (6.5) is obtained by substituting into

Equation (6.4) the numerical values of the luminosity coefficients which apply to the reference stimuli used for colour television.

We will next consider some of the reasons for choosing to transmit luminance and colour difference signals and some of the consequences of this choice.

Compatibility

The choice of luminance and colour difference signals is consistent with compatibility because

(a) The luminance signal E_Y is modulated onto an r.f. carrier in the same way as the video signal for monochrome television.

(b) The form of modulation used for the colour difference signals ensures that they have a negligible effect on the picture displayed by a monochrome receiver. This will be discussed further in Section 6.8.

Reverse compatibility

As explained in Section 6.2.5, the chrominance values for a colour source are proportional to $E_R - E_Y$, $E_G - E_Y$, and $E_B - E_Y$. This is why the colour difference signals are also called **chrominance signals.**

The chrominance of a monochrome grey source is zero so that a black and white scene produces zero colour difference signals. This is the key to reverse compatibility. A monochrome transmission contains no colour difference signals. A colour set receiving such a transmission behaves as if the colour difference signals were zero; that is $E_R - E_Y = 0$ and $E_G - E_Y = 0$, giving $E_R = E_Y$ and $E_G = E_Y$ directly. The fact that E_B is also equal to E_Y follows, less directly, from $E_Y = lE_R + mE_G + nE_B$ (Equation (6.4)) with $l + m + n = 1$ (Equation (6.3)) and $E_R = E_B = E_Y$.

A monochrome transmission therefore produces pictures with equal values of E_R, E_G and E_B, that is equal tristimulus values, corresponding to reference white for the system. Also the luminance of the picture produced by the tube is proportional to $lE_R + mE_G + nE_B = (l + m + n)E_Y = E_Y$, which is the luminance signal for the monochrome transmission.

A colour set receiving a monochrome transmission therefore produces a black and white (grey) picture whose chromaticity corresponds to that of the reference white and whose luminance corresponds to the luminance of the scene being televised.

Relation between the colour difference signals

Any colour has three chrominance values and therefore corresponds to three colour difference signals $(E_R - E_Y)$, $(E_G - E_Y)$, and $(E_B - E_Y)$. Only two need to be transmitted because, rather like chromaticity coordinates, the third can be calculated if the other two are known. Using Equations (6.3) $(l + m + n = 1)$ and (6.4) $(E_Y = lE_R + mE_G + nE_B)$ we can write

218

$$E_Y = (l + m + n)E_Y = lE_R + mE_G + nE_B$$

so that

$$l(E_R - E_Y) + m(E_G - E_Y) + n(E_B - E_Y) = 0$$

which is a linear relation between the three colour difference signals. Thus, if we know $(E_R - E_Y)$ and $(E_B - E_Y)$, we can obtain $(E_G - E_Y)$ in the form

$$(E_G - E_Y) = -\frac{l}{m}(E_R - E_Y) - \frac{n}{m}(E_B - E_Y)$$
$$= -0{\cdot}51(E_R - E_Y) - 0{\cdot}19(E_B - E_Y) \qquad (6.6)$$

The last line was obtained by substituting the standard colour television values $l = 0{\cdot}30$, $m = 0{\cdot}59$ and $n = 0{\cdot}11$ for the luminosity coefficients.

It can be shown that, for any colour, the green colour difference signal $(E_G - E_Y)$ is smaller than, or equal to, the red and blue colour difference signals $(E_R - E_Y)$ and $(E_G - E_Y)$ (essentially because the green luminosity coefficient is greater than the other two). This result shows that choosing a pair of signals which includes the green difference signal would produce signals more susceptible to interference than those obtained by choosing the blue and red difference signals. This is a reason why the red and blue colour difference signals were chosen for transmission.

A matrix circuit
Figure 6.12 shows the principle of a receiver matrix circuit used to obtain the gun voltages $-E_R$, $-E_G$ and $-E_B$ from the luminance signal E_Y and the two colour difference signals $(E_R - E_Y)$ and $(E_B - E_Y)$. The circuit consists of four operational amplifiers acting as inverting adders. The summing resistors connected to the input of amplifier 1 are chosen to give a weighted sum of the inputs according to Equation (6.6). The output of this amplifier is therefore $-[0{\cdot}51(E_R - E_Y) + 0{\cdot}19(E_B - E_Y)]$, which is equal to $(E_G - E_Y)$ (the minus sign outside the bracket is due to the use of an inverting amplifier). The resistors of the other three amplifiers are chosen to give simple unweighted sums. The output of amplifier 2 is $-[(E_R - E_Y) + E_Y] = -E_R$, and, similarly, the outputs of amplifiers 3 and 4 are $-E_G$ and $-E_B$ respectively. Figure 6.12 is intended to show the principle of a matrix circuit. The functions of amplifiers 1 and 3 could be served by a single three-input amplifier in a practical circuit.

Noise and the constant luminance principle
The choice of transmitted signals satisfies the requirement of the constant luminance principle that the luminance signal should not be affected by any noise superimposed on the chrominance signal. This can be shown as follows.

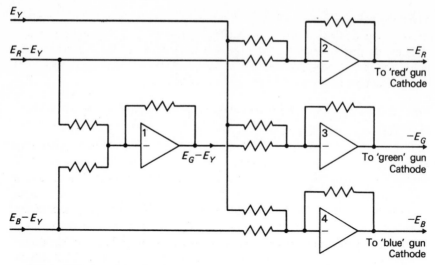

Figure 6.12 A matrix circuit. Input voltages E_Y, $E_R - E_Y$ and $E_B - E_Y$ are transformed into picture tube cathode voltages $-E_R$, $-E_G$ and $-E_B$ using four operational amplifiers as inverting adders

Imagine that the luminance signal E_Y is transmitted and received without being affected by noise, but that the colour difference signals $(E_R - E_Y)$ and $(E_B - E_Y)$ are received as $(E_R - E_Y) + N_1$ and $(E_B - E_Y) + N_2$, where N_1 and N_2 represent additional noise voltages generated in the system.

The grid-cathode voltages of the three cathode ray tube guns, obtained by combining the luminance and colour difference signals in the receiver matrix circuit, are:

$$[(E_R - E_Y) + N_1] + E_Y = E_R + N_1 \text{ for the red gun}$$

$$[(E_B - E_Y) + N_2] + E_Y = E_B + N_2 \text{ for the blue gun}$$

and, using Equation (6.6)

$$-\frac{l}{m}[(E_R - E_Y) + N_1] - \frac{n}{m}[(E_B - E_Y) + N_2] + E_Y$$

for the green gun.

Thus all three signals are affected by noise. The luminance obtained on the screen is proportional to the sum of these three signals weighted according to the luminosity coefficients l, m and n. The screen luminance is therefore proportional to

$$l(E_R + N_1) + m\left[-\frac{l}{m}\{(E_R - E_Y) + N_1\} - \frac{n}{m}\{(E_B - E_Y) + N_2\} + E_Y\right]$$
$$+ n(E_B + N_2)$$

The noise voltages cancel and this expression simply reduces to $lE_Y + mE_Y + nE_Y$ which is equal to E_Y because $l + m + n = 1$ by Equation (6.3).

Thus, despite the fact that the voltages applied to all three guns include noise, the luminance produced on the screen is free from noise effects. It is only the colour of the picture which is affected.

6.4.3 Gamma corrections and their effects

Gamma correction is carried out at the picture source, as explained in Section 5.12. A correction corresponding to a picture tube gamma of 2·8 is normally specified but the gamma for picture tubes ranges between about 2·2 and 3·3. The correction actually applied is under the control of the studio technical staff. It may be chosen on the basis of a subjective judgment of the quality of a picture viewed on a monitor, and could correspond to a receiver picture tube gamma as low as 1·8. Such a correction for a tube with an actual gamma of 2·7, say, would give an overall, or system, gamma of $2\cdot7/1\cdot8 = 1\cdot5$. A system gamma greater than unity results in most colours being reproduced with a greater saturation than they had in the original scene.

Whatever its magnitude, the gamma correction process involves certain choices and compromises. The luminance signal from a three-tube camera could be formed by combining the three uncorrected camera outputs in a matrix circuit to give $E_Y = 0\cdot30\,E_R + 0\cdot59\,E_G + 0\cdot11\,E_B$, and this signal could be passed through a non-linear network to give an output $E_Y{}^{\gamma_c}$. The red and blue camera outputs could also be passed through similar networks and then combined with $E_Y{}^{\gamma_c}$ to give ideal colour difference signals $E_R{}^{\gamma_c} - E_Y{}^{\gamma_c}$ and $E_B{}^{\gamma_c} - E_Y{}^{\gamma_c}$. The process is illustrated in Figure 6.13(a). The objection to this arrangement is that non-linear networks would be needed in the receiver in order to recover the green difference signal, as can be verified by writing down the appropriate sums, starting from $E_Y{}^{\gamma_c}$ and the gamma corrected red and blue colour difference signals. In order to avoid this difficulty the compromise arrangement of Figure 6.13(b) is used.

Gamma corrections are applied directly to the outputs of the three camera tubes and, instead of $E_Y{}^{\gamma_c}$, the quantity

$$E'_Y = 0\cdot30\,E_R{}^{\gamma_c} + 0\cdot59\,E_G{}^{\gamma_c} + 0\cdot11\,E_B{}^{\gamma_c} \tag{6.7a}$$

is formed by taking the sum of the gamma corrected colour voltages weighted according to the luminosity coefficients of the three primary colours.

It is customary to use E'_Y for the quantity defined in Equation (6.7a) and to use E'_R for $E_R{}^{\gamma_c}$, E'_B for $E_B{}^{\gamma_c}$ and E'_G for $E_G{}^{\gamma_c}$. Thus Equation (6.7a) is usually written

$$E'_Y = 0\cdot30\,E'_R + 0\cdot59\,E'_G + 0\cdot11\,E'_B \tag{6.7b}$$

221

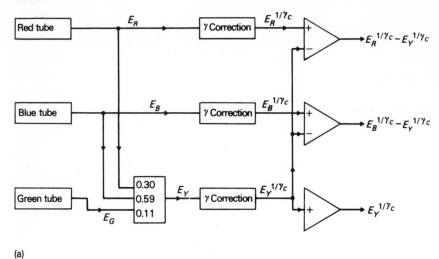

(a)

CAMERA

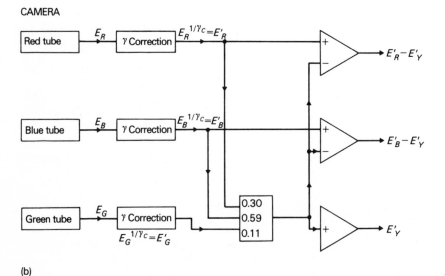

(b)

Figure 6.13 Camera output gamma correction circuits. (a) The three camera signals are combined in a matrix to form E_Y before gamma correction to produce $E_Y^{1/\gamma}$. (b) The three camera signals are gamma corrected before being combined to produce E_Y'

but remember that, in general

$$E'_Y \neq E_Y{}^{\gamma c}$$

because for any three numbers a, b, c, $(a + b + c)^\gamma \neq a^\gamma + b^\gamma + c^\gamma$, unless two of the numbers are zero.

Returning to Figure 6.13(b), the red and blue colour difference signals are formed by subtracting E'_Y from E'_R and E'_B respectively. This arrangement makes the recovery of green colour difference signal in the receiver much simpler. Using Equation (6.5) and the same argument as that of Section 6.4.2 (Equation (6.6)) we see that

$$E'_G - E'_Y = -0{\cdot}51(E'_R - E'_Y) - 0{\cdot}19(E'_B - E'_Y) \tag{6.8}$$

So the green colour difference signal, and hence E'_G, can be obtained from a simple, linear, matrix circuit of the type shown in Figure 6.12.

The use of E'_Y instead of $E_Y{}^{\gamma c}$ has several disadvantages which will now be considered. They involve departure from the constant luminance principle. However, the designers of existing television systems have concluded that these disadvantages are more than outweighed by the simplification in the receiver circuits.

Loss of compatibility

The luminance signal for a monochrome receiver tuned to a colour transmission is E'_Y. The 'correct' signal is $E_Y{}^{\gamma c}$ and the difference between them represents a loss of compatibility.

In general the form of gamma correction which is used causes a monochrome receiver operating on a colour signal to produce a luminance level lower than the correct one; except if the signal corresponds to grey, in which case the three colour signals are equal. So with $E_R = E_B = E_G = E$, say

$$E_Y{}^{\gamma c} = (0{\cdot}30E + 0{\cdot}59E + 0{\cdot}11E)^{\gamma c} = E^{\gamma c}$$

and

$$E'_Y = 0{\cdot}30E^{\gamma c} + 0{\cdot}59E^{\gamma c} + 0{\cdot}11E^{\gamma c} = E^{\gamma c}$$

since

$$0{\cdot}30 + 0{\cdot}59 + 0{\cdot}11 = 1$$

Hence in this case the two signals are equal and transmission of E'_Y will produce the correct luminance.

The loss of compatibility is not as bad as might appear at first sight. The error is smaller for desaturated colours than for saturated ones. There is also

223

an unrelated effect which acts in the opposite direction and tends to reduce the error. The chrominance signal is modulated onto a carrier, as will be explained in Section 6.5. All you need to know for the present is that, for any uniform colour other than grey, the chrominance signal will consist of a sinusoid in the upper part of the video band. This sinusoid will produce a set of bright and dark patches on the screen of a monochrome receiver. If the receiver picture tube had a gamma of one, the increase of luminance due to the bright patches would just balance the decrease due to the dark patches and there would, on average, be no change in luminance. However, since gamma is greater than one in practice, the increase in luminance in the bright patches, corresponding to the maxima of the sinusoids, is greater than the decrease corresponding to minima. The non-linearity of the tube response produces an increase in the mean signal level which results in an increase in average luminance.

Effect on noise in colour receivers

One consequence of the departure from the constant luminance principle is that the picture luminance will no longer be immune to noise which affects the colour difference signals. (The change from $E_Y{}^{\gamma_c}$ to E'_Y prevents cancellation of the effect on picture tube luminance of noise in the colour difference signals.) However, here again, the resulting degradation is considered to be acceptable.

Effect on the luminance of colour signals

At first sight, referring to Figure 6.12, the effects of gamma corrections simply result in the red, green and blue gun signals being modified to $-E'_R$, $-E'_G$ and $-E'_B$, respectively. These signals produce the correct luminance for the three primary colour pictures and hence the viewer perceives a combined picture with the correct luminance. However, whereas the luminance signal E'_Y occupies a bandwidth of the order of 5 MHz, the chrominance signals $(E'_R - E'_Y)$ and $(E'_B - E'_Y)$ are limited to a bandwidth of the order of 1 MHz, because, as mentioned in Section 6.4.1 the eye can perceive less chrominance than luminance detail. The reduction of chrominance bandwidth removes detail which cannot be seen in any case, while at the same time increasing compatibility, because, as will be explained in Section 6.8, it allows the chrominance signal to be modulated onto a sub-carrier so as to occupy a part of the video spectrum where the luminance signal amplitudes are generally small.

Looking at Figure 6.12, the signal for the red gun is obtained as the sum of two signals E'_Y and $E'_R - E'_Y$. The summation is carried out in amplifier 2, and the result is $-E'_R$, because an inverting amplifier is used. This red signal is obtained through the cancellation of the luminance signal E'_Y and the $-E'_Y$ part of the colour difference signal. This cancellation only occurs in the frequency band occupied by the colour difference signal, that is in the chrominance band. In the rest of the video band the colour difference signal is negligibly small, but the luminance signal is not, so that the red gun signal becomes $-E'_Y$, instead of $-E'_R$. The same thing happens for the other two guns. Their signals also become equal to $-E'_Y$. The observed luminance is the combined effect of the three primary luminances, that is $0{\cdot}30\,(E'_Y)^\gamma + 0{\cdot}59\,(E'_Y)^\gamma + 0{\cdot}11(E'_Y)^\gamma = (E'_Y)^\gamma$. Thus, outside the chrominance band, the luminance is in error to the extent that $E'_Y \neq E_Y{}^{\gamma_c}$. (Remember that, from Section 5.12, the observed luminance is required to be proportional to the camera output voltage raised to the power $\gamma_c\gamma$.)

User tests have shown that the departure from complete compatibility, the effect of chrominance noise on picture luminance, and the luminance errors outside the chrominance band are acceptable prices to pay for the receiver simplification achieved by using E'_Y instead of $E_Y{}^{\gamma_c}$ for the gamma corrected luminance signal.

Four-tube cameras
The luminance signal in a three-tube camera is formed by combining the outputs of all three tubes according to Equation (6.5). The luminance information in the received picture is therefore effectively built up by superposing three pictures. If these three pictures do not coincide accurately enough, the fine luminance detail of the scene will appear blurred on the receiver screen. The registration of the three camera tubes, that is the degree of coincidence between the three pictures they produce, and, hence, the mechanical alignment of the optical systems for the three tubes, is therefore very critical. This is less of a problem in four-tube cameras which use one tube for the broad band luminance signal carrying most of the picture detail, and three tubes for the relatively narrow band colour difference signals which carry much less picture detail, so that their registration is much less critical.

Three-tube camera circuits have been used in this section to illustrate the type of problems caused by gamma corrections in colour television. Related problems arise in the use of four-tube cameras. Gamma corrections with four-tube cameras are discussed in References 64 and 66. A four-tube camera and its gamma correction circuits are described in Reference 95.

6.5 QUADRATURE MODULATION

The form of modulation used for the chrominance signal in the NTSC and PAL systems is described in this section.

Because of the need for compatibility, the luminance signal takes the same form as in monochrome systems. The chrominance (colour difference) signals occupy less than the full video bandwidth, and there is room, within this bandwidth, to modulate them onto sinusoidal waveforms called sub-carriers. If two subcarriers of different frequencies were used, these two frequencies would produce visible beat patterns. To reduce this, both chrominance signals are modulated onto subcarriers of the same frequency, but differing in phase by 90°. This is known as **quadrature modulation.** Even with a single frequency a beat pattern is produced on monochrome receivers, but it is much less visible than the set of patterns which would be formed by subcarriers of different frequencies.

As will be explained in Section 6.6, the colour difference signals $(E'_B - E'_Y)$ and $(E'_R - E'_Y)$ are not used directly. This is in order to avoid overloading the transmitter. They are first reduced in magnitude to give signals U and V, with $U = 0\cdot877\,(E'_R - E'_Y)$ and $V = 0\cdot493\,(E'_B - E'_Y)$. U and V are time-varying signals with Fourier components limited to less than the video band.

U modulates the carrier $E_{sc} \cos \omega_{sc} t$ in a balanced modulator giving a double sideband suppressed carrier output $UE_{sc} \cos \omega_{sc} t$. V similarly modulates the quadrature carrier $E_{sc} \sin \omega_{sc} t$ to give $VE_{sc} \sin \omega_{sc} t$. The resulting modulated chrominance signal obtained by combining the two is

$$UE_{sc} \cos \omega_{sc} t + VE_{sc} \sin \omega_{sc} t$$

A block diagram of an encoder, that is the part of a transmitter used to produce this signal, is shown in Figure 6.14. The bandpass filters are used to limit the bandwidth of the modulated signals.

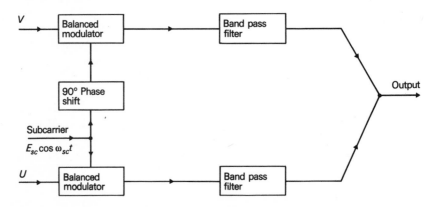

Figure 6.14 NTSC chrominance modulator (encoder)

At the receiver synchronous detectors are used for demodulation. A local subcarrier $F_{sc} \cos (\omega_{sc} t + \theta)$ is generated. It has amplitude F_{sc} and phase θ relative to the transmitter carrier $E_{sc} \cos \omega_{sc} t$. (As will be explained in Section 6.7 the local carrier phase and frequency are controlled by a synchronising signal, the colour burst, transmitted during the line blanking period.) With an input signal proportional to $UE_{sc} \cos \omega_{sc} t + VE_{sc} \sin \omega_{sc} t$, the output of a synchronous detector will be of the form

$E_{sc} F_{se}[U\cos \omega_{se} t \cos (\omega_{se} t + \theta) + V \sin \omega_{sc} t \cos (\omega_{sc} t + \theta)]$

$= \frac{1}{2} E_{sc} F_{se}[U\{\cos (2\omega_{se} t + \theta) + \cos \theta\} + V \{\sin (2\omega_{sc} t + \theta) - \sin \theta\}]$

If the high frequency $(2\omega_{sc})$ terms are removed using a low pass filter, the final output will be proportional to

$$\frac{1}{2} UE_{sc} F_{sc} \cos \theta - \frac{1}{2} VE_{sc} F_{sc} \sin \theta$$

so that if $\theta = 0$, the output is proportional to U and if $\theta = \pi/2$ the output is proportional to V. For convenience, we will assume, from here on, that gains and signal levels have been set in the receiver so that $\frac{1}{2} E_{sc} F_{sc} = 1$ and refer to U and V directly as the amplitudes of the subcarrier components. We can therefore write

$$\text{demodulator output} = U \cos \theta - V \sin \theta \qquad (6.9)$$

Two synchronous demodulators are used in the decoder circuit of a receiver. The block diagram of a decoder is shown in Figure 6.15. An extra phase shift φ is introduced. If $\varphi = 1$ the decoder outputs are U and V, but some systems use non-zero values of φ to give the outputs proportional to $U \cos \varphi - V \sin \varphi$ and $U \cos (\varphi + \pi/2) - V \sin (\varphi + \pi/2) = -(U \sin \varphi + V \cos \varphi)$ as shown in the figure. This is discussed further in Section 6.9.

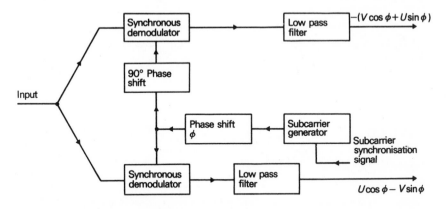

Figure 6.15 NTSC chrominance demodulator (decoder)

In considering the choice of chrominance signals and the significance of these choices in terms of picture defects, it is useful to use phasor diagrams for the modulated chrominance signal. We will now see how this can be done in the case of quadrature modulation.

First consider a single Fourier component of signal U, which can be written $E_U \cos \omega t$, used to amplitude modulate subcarrier $E_{sc} \cos \omega_{sc} t$. The appropriate phasor diagram is shown in Figure 6.16(a). This is exactly the same situation as that of Figure 5.26 of Section 5.9.1. We can resolve the two components along the direction of the subcarrier, to give $E_U \cos \omega t$; whereas normal to it, the two components cancel. Now, because we are using suppressed carrier double sideband modulation, the final result is simply a phasor $E_U \cos \omega t$ along the original direction of the carrier as shown in Figure 6.16(b). If we consider signal U to be made up of a sum of Fourier components of the form $E_U \cos \omega t$ we can express the whole signal as a vector U along the direction of the subcarrier (Figure 6.16(c)).

We can consider signal V similarly. It is modulated onto a carrier 90° out of phase with that of the U signal. This gives the same diagram as Figure 6.16(c) rotated through 90° and, by the same argument, leads to a vector V at 90° to the direction of the subcarrier, and hence of U, so that the combined phasor diagram is simply Figure 6.16(d).

In Figure 6.16(d) the lengths (magnitudes) of U and V vary with time for a general signal. They cannot be thought of as phasors as they are not, in general, sinusoids, but they can be thought of as sums of sinusoidal Fourier components. When dealing with signals U and V we will often be concerned with areas of uniform colour, in which case U and V will behave as zero

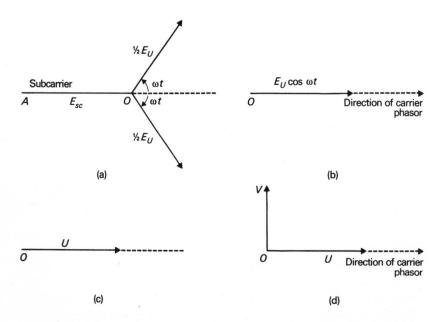

Figure 6.16 (a) Amplitude modulated subcarrier. (b) Envelope demodulation of a sinusoidal signal which has been transmitted by double sideband suppressed carrier modulation. (c) Envelope demodulation of a general signal U. (d) Envelope demodulation of two signals U and V with the subcarriers for the two signals in phase quadrature

frequency phasors ($\omega = 0$) and will not change with time. The diagram of Figure 6.16(d) represents the electrical signal $U \cos \omega_{sc} t + V \cos (\omega_{sc} + \frac{1}{2}\pi)$ $= U \cos \omega_{sc} t - V \sin \omega_{sc} t$. So U and V are proportional to the double sideband suppressed carrier modulation on two carriers of the same frequency differing by 90° in phase.

Thus, for regions of uniform colour (U and V constant), the two modulated signals have the same frequency, but differ in phase by 90°. As shown in Figure 6.17 they can be represented by a single phasor of amplitude S and relative phase φ, where

$$S = (U^2 + V^2)^{1/2}$$

and (6.10)

$$\varphi = \tan^{-1} V/U$$

the phase being measured with respect to the subcarrier which is modulated by the U signal. This reference direction will be called the U **axis.** The U signal is proportional to the blue colour difference signal $E'_B - E'_Y$ in the NTSC and PAL system.

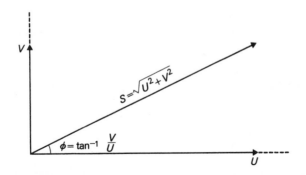

Figure 6.17 Phasor representation of two constant signals, representing a uniform colour, modulating two subcarriers in phase quadrature

6.5.1 Quadrature distortion

The chrominance signals are band limited by being passed through a filter. If the passband of the filter is not symmetrical there will be some quadrature distortion. A similar effect was discussed in Section 5.9 when considering vestigial sideband demodulation, except that the carrier is suppressed in the present case. Consider a Fourier component $E_U \cos \omega t$ of signal U. If it is modulated onto a subcarrier of unit amplitude, using double sideband suppressed carrier modulation, it will contribute $\frac{1}{2}E_U \cos (\omega_{sc} + \omega)t$ and $\frac{1}{2}E_U \cos (\omega_{sc} - \omega)t$ to the upper and lower sidebands respectively. On demodulation a signal $E_U \cos \omega t$ will be recovered. This corresponds to Figure 6.16(b). It is the phasor which, before demodulation, represents the resultant of the two sidebands along the direction that is in phase with the (suppressed) subcarrier. If, however, the modulated signal is passed through an asymmetrical filter which attenuates the upper sideband, but not the lower one, the signal reaching the demodulator will be of the form

$$\tfrac{1}{2}E_U \cos (\omega_{sc} - \omega)t + \tfrac{1}{2}\,\delta E_U \cos (\omega_{sc} + \omega)t$$

with $\delta < 1$. This is shown in the phasor diagram of Figure 6.18(a). When the phasors are resolved parallel and normal (in quadrature) to the subcarrier (Figure 6.18(b)) the quadrature components no longer cancel. Thus, after synchronous demodulation, there is a signal of amplitude $\tfrac{1}{2}(1 + \delta)E_U$, instead of E_U. There is also a spurious contribution of amplitude $\tfrac{1}{2}(1 - \delta)E_U$ to the output of the second synchronous demodulator which uses a local

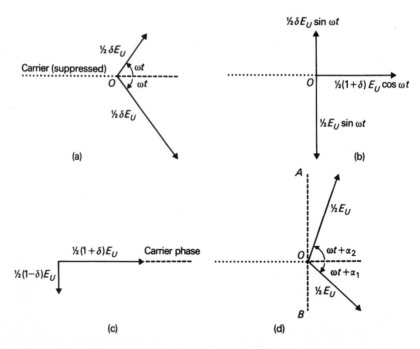

Figure 6.18 (a) Double sideband suppressed carrier signal which has different upper and lower sideband amplitudes as the result of filtering with a filter which attenuates the upper sideband more than the lower one. (b) Phasors of Figure 6.18(a) resolved in phase and in quadrature with the carrier. (c) Envelope demodulation produces signal amplitude $\tfrac{1}{2}(1 + \delta)E_U$ if the local subcarrier is in phase with the original subcarrier and a signal of amplitude $\tfrac{1}{2}(1 - \delta)E_V$ if it is in quadrature. (d) Phasor diagram for the effect of a filter which produces phase shifts of different magnitudes for the upper and lower sidebands of a d.s.b.s.c. signal

subcarrier in phase quadrature with the subcarrier for the first demodulator, as can be seen from Figure 6.18(c). This second signal appears as a contribution to the V chrominance signal and cannot be distinguished from it. Asymmetric filtering therefore causes mixing of the U and V signals. This is called **phase quadrature crosstalk.**

A similar effect, leading to crosstalk, will occur if the sideband amplitudes are unchanged, but if their phases are shifted through angles α_1 and α_2 which are not equal and opposite with respect to the carrier. The situation is shown in Figure 6.18(d). The phasors both have length $\frac{1}{2}E_U$, but their quadrature components OA and OB are not equal and opposite so that they do not cancel. This form of distortion is produced when the modulated subcarrier is passed through a circuit whose group delay varies with frequency. It is therefore important to keep this variation to a minimum in the chrominance circuits.

6.6 COLOUR BAR SIGNALS

Colour bar signals are used in testing the performance of receivers, but the reasons why they are discussed here are firstly to familiarise you with chrominance signals, secondly to show how these signals can be represented on a vector diagram, thirdly to introduce the weighting factors which are used in the NTSC and PAL systems, and fourthly to introduce the E'_I and E'_Q signals which are used in the NTSC system.

Consider the red, green and blue colour signals E'_R, E'_G and E'_B of Figure 6.19 and imagine that they are combined to form a composite video signal, consisting of a luminance signal $E'_Y = 0.30E'_R + 0.59E'_G + 0.11E'_B$, according to Equation (6.7), and two colour difference signals $E'_R - E'_Y$ and $E'_B - E'_Y$ which provide the chrominance information.

During each line period there are eight different combinations of signals, as can be seen from Figure 6.19. The signals repeat on each line. They therefore build up a picture consisting of eight vertical stripes, called **colour bars.** Each bar is uniformly of one colour but the colours differ from bar to bar. If the colour difference signals are quadrature modulated directly onto the suppressed colour-subcarriers, they will produce a sinusoid at the colour-subcarrier frequency with amplitude $E'_C = [(E'_B - E'_Y)^2 + (E'_R - E'_Y)^2]^{1/2}$ for the duration of each bar, that is for one eighth of an active line period. These quantities together with others, dealt with below, are shown in Table 6.1.

Since we are mainly concerned, in this section, with the luminance and chrominance parts of the composite video signal, the voltages have been normalised so that the luminance signal ranges from 0 to 1 V, with the black level at 0 and the white level at $+1$ V. Now for most systems the range from black to white is approximately 70 per cent of the range from sync level to white; so with the white level at 1 V we must take the sync level as 30/70 V

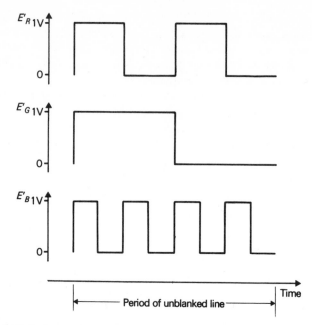

Figure 6.19 Red, green and blue signals used to produce a colour bar picture

below black level, that is at -0.43 V. This normalisation saves a good deal of arithmetic because the three signals E'_R, E'_G, and E'_B only take the two values 0 or 1 in our signals.

The colours of the bars are listed in Table 6.1. They can be deduced from Figure 6.3. For instance, the first bar with $E'_R = E'_G = E'_B = 1$ V is white, and the second with $E'_R = E'_G = 1$ V and $E'_B = 0$ V is a mixture of red and green. It is therefore yellow.

The sequence of signals has been chosen to give a decreasing luminance from one bar to the next. You can verify this by checking the values of E'_Y in Table 6.1. These values are also plotted in Figure 6.20(a). They remain constant over the duration of each bar. The modulated chrominance signals vary sinusoidally at the chrominance subcarrier frequency, but, as will be explained in Section 6.8, this frequency is large compared with the line frequency. The chrominance part of the signal therefore looks like a blur when displayed on an oscilloscope swept at the line frequency. Figure 6.20 is effectively three such displays, and Figure 6.20(b) shows the modulated chrominance signal for each bar as a blurred rectangle extending on either side of the time axis by $\pm E'_C$. The composite video signal, obtained by combining the luminance signal E'_Y and the modulated chrominance signal of amplitude E'_C, is shown in Figure 6.20(c). The limiting values of this signal are $E'_Y \pm E'_C$. These are also listed in Table 6.1. You may like to verify a few entries in this table and the corresponding parts of Figure 6.20, to make sure you understand how they are obtained.

232

Table 6.1 *Colour bar signals*

Colour of bar	E'_R	E'_G	E'_B	E'_Y	$E'_B - E'_Y$	$E'_R - E'_Y$	E'_C	$E'_Y + E'_C$	$E'_Y - E'_C$
White	1	1	1	1·00	0	0	0	1·00	1·00
Yellow	1	1	0	0·89	−0·89	0·11	0·90	1·79	−0·01
Cyan	0	1	1	0·70	0·30	−0·70	0·76	1·46	−0·06
Green	0	1	0	0·59	−0·59	−0·59	0·83	1·42	−0·24
Magenta	1	0	1	0·41	0·59	0·59	0·83	1·24	−0·42
Red	1	0	0	0·30	−0·30	0·70	0·76	1·06	−0·46
Blue	0	0	1	0·11	0·89	−0·11	0·90	1·01	−0·79
Black	0	0	0	0	0	0	0	0	0

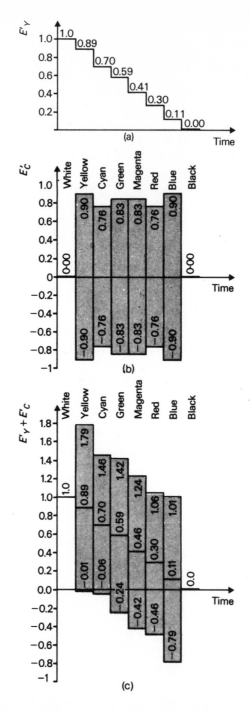

Figure 6.20 Video signals for colour bar picture during the active part of a line period. (a) Luminance signal. (b) Chrominance signal. (c) Luminance and chrominance signals combined

6.6.1 The weighted chrominance signals

We have normalised our signals to $+1$ V corresponding to peak white with a blanking level of 0 V and a sync level of -0.43; but the composite signal varies well beyond these limits, and this would overload the transmitter. In order to reduce this overload, the magnitude of the colour difference signals is reduced in relation to the luminance signal. This is done by using $0.877 (E'_R - E'_Y)$ instead of $E'_R - E'_Y$ and $0.493 (E'_B - E'_Y)$ instead of $E'_B - E'_Y$ to modulate the chrominance subcarrier. The factors 0.877 and 0.493 are called **weighting factors.**

The colour bar signals represent an extreme case because they include the largest composite video signals which can arise for any picture. The weighting factors which are adequate for the colour bar signals are therefore adequate for any signal which may be transmitted. These factors are used for all transmitted chrominance signals.

The weighted chrominance signals are

$$E'_U = 0.877(E'_R - E'_Y) \text{ and } E'_V = 0.493(E'_B - E'_Y) \qquad (6.11)$$

One sometimes finds the weighting factors written as $1/1.14$ and $1/2.03$.

Using the weighted chrominance signals, the peak values of the composite video signal are

$$E'_Y \pm (E'^2_U + E'^2_V)^{1/2}$$

because E'_U and E'_V appear as quadrature components of the modulated subcarrier.

These peak values can be calculated for the colour bar signal, using Table 6.1 for E'_Y, $E'_R - E'_Y$ and $E'_B - E'_Y$. This is left as an exercise. The results are plotted in Figure 6.21. It can be seen from this figure that the variation of the normalised signal has been limited to the range -0.33 to $+1.33$ V.

The choice of weighting factors represents a compromise between the effect of overloading and of noise. Maximum values of the combined signal, corresponding to high luminance, highly saturated colours, only occur for limited periods (of the order of 1 per cent of the time) in average television transmissions. A certain amount of overloading is therefore allowed, because this leads to higher signal voltages overall, so that the signals, and in particular lower amplitude ones, are less affected by noise than they would be if weighting factors were chosen to keep the combined signal between our normalised values 0 and 1 V at all times. There is some overloading between 1 and 1.33 V. The signal in the sync region between 0 and -0.33 V does not cause problems because each excursion of the video signal into the sync region takes place during a small fraction of the period of the chrominance subcarrier. This is too short a time to cause spurious operation of the line scan generator.

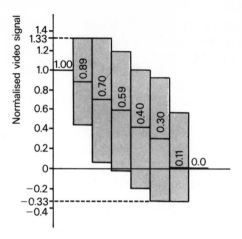

Figure 6.21 Combined video signal using the weighted chrominance signals of Equation (6.11)

6.6.2 Phasor representation of the colour bar signals

During each colour bar, the modulated chrominance signal is a sinusoid at the colour subcarrier frequency. It has components E'_U and E'_V given by Equation (6.11), where the values of E'_Y, E'_R and E'_B are given in Table 6.1. This signal can therefore be represented as a phasor with amplitude

$$S = (E'_U{}^2 + E'_V{}^2)^{1/2}$$

and making an angle φ with the U axis, where

$$\varphi = \tan^{-1} E'_V / E'_U$$

The phasors for the colour bar signals are shown in Figure 6.22. The tip of each phasor has cartesian coordinates E'_U and E'_V and this provides a simple way of drawing the diagram using Equation (6.11) and Table 6.1.

Each bar corresponds to a phasor terminating on one of the vertices of a polygon. It can be shown that any colour which is less saturated, or has less than maximum luminance, corresponds to a phasor terminating inside the polygon.

We have been dealing, up to now, with gamma corrected signals. However, the signals E'_R, E'_B and E'_G which are gamma corrected in the transmitter have all been 0 or 1 V, so that there was no numerical difference between the corrected and uncorrected signals. We will now be considering values and results that will be simpler to analyse if we do not work with gamma corrected signals, or assume a transmitted gamma of value one.

Gamma correction modifies and complicates the details but not the

236

underlying principles of the ideas which will now be discussed, and it is the principles which are the concern of this book.

If phasors are plotted for various colours and luminances on a diagram such as that of Figure 6.22, the angles φ, which the phasors make with the U-axis are found to depend on the hue of the colour, and their lengths S depend on both the luminance and saturation. But if the luminance were fixed, an increase in S would represent an increase in saturation. Hence diagrams such as Figure 6.22 enable us to estimate the visual result of system defects leading to changes in subcarrier phase or amplitude.

6.6.3 The E_I and E_Q signals in the NTSC system

Visual sensitivity to chromaticity changes depends on the direction of these changes. This was indicated in Figure 6.8 of Section 6.2.7. The eye is less sensitive to chromaticity changes along the direction of the major axes of the ellipses shown in that figure than it is along the direction of the minor axes.

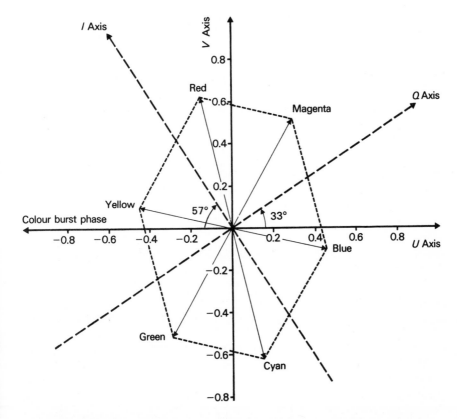

Figure 6.22 Phasor diagram showing the colour bar chrominance signals corresponding to Figure 6.19, the I and Q axes, and the phase of the colour burst for the NTSC system.

Thus, if the two chrominance signals could represent changes along these two directions, one signal would require less bandwidth than the other in order to convey what would be perceived as the same amount of chromaticity detail. This principle is used in the NTSC system.

The chrominance signals E_U and E_V depend both on luminance which is proportional to E_Y, and on chromaticity, so they cannot be represented directly on a chromaticity diagram. But the quantities E_U/E_Y and E_V/E_Y depend on chromaticity alone. Contours of constant E_U/E_Y and E_V/E_Y are plotted on a CIE chromaticity diagram in Figure 6.23. They form an array of lines which are nearly at right angles. Notice the direction of the two sets of

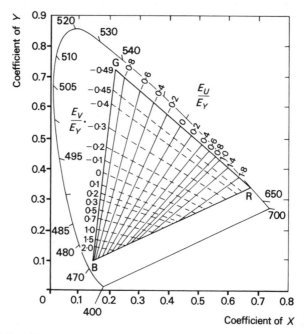

Figure 6.23 Lines of constant E_U/E_Y and E_V/E_Y on a CIE chromaticity diagram

lines. They do not lie along the directions of maximum and minimum visual sensitivity of Figure 6.8, but it is possible to choose axes that do so, forming the following signals

$$E_Q = E_V \sin 33° + E_U \cos 33°$$
$$E_I = E_V \cos 33° - E_U \sin 33° \qquad (6.12)$$

The lines of constant E_Q/E_Y and E_I/E_Y on a chromaticity diagram are shown in Figure 6.24. It can be seen from this that lines of constant E_I/E_Y lie roughly along a direction such that changes in E_I/E_Y, for constant

238

E_Q/E_Y, are in the direction of maximum visual sensitivity to chromaticity changes. Similarly, changes of E_Q/E_Y, for constant E_I/E_Y, take place along the direction of minimum sensitivity.

Thus by choosing E_I and E_Q as chrominance signals it is possible to use narrower bandwidths for E_Q than for E_I. In the North American 525-line NTSC system, the E_Q signal is specified as having a 2 dB bandwidth of 400 kHz and a 6 dB bandwidth of 500 kHz. The E_I signal is specified as having a 2 dB bandwidth of 1·3 MHz and a 20 dB bandwidth of 3·6 MHz. It is more usual to specify just one bandwidth, normally at 3 dB. The use of two bandwidths for each signal provides more precise information about the

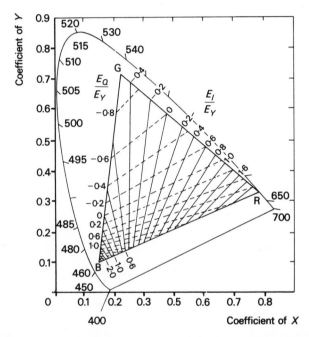

Figure 6.24 Lines of constant E_Q/E_Y and E_I/E_Y on a CIE chromaticity diagram

required characteristics of the filters which are used for band-limiting the signals.

Equation (6.12) implies that there is an angle of 33° between the U, V and the Q,I axes. This is shown in Figure 6.22.

Figures 6.23 and 6.24 are for the $\gamma_t = 1$ case. The transmitted gamma is typically chosen to be 2·2, in which case we must have E'_Q, E'_I, E'_U and E'_V instead of E_Q, E_I, E_U and E_V in Equation (6.12). When γ_t is not unity, contours of constant E'_U, E'_V and constant E'_Q, E'_I become curved. But their general direction still justifies the above choice of Q and I axes.

6.7 THE NTSC COLOUR BURST

The chrominance information is carried in the amplitude and phase of the modulated subcarrier. The phase is relative to a standard signal. At the transmitter this standard signal is the unmodulated subcarrier. At the receiver it is a locally generated signal in phase with the transmitter subcarrier and at the same frequency. This local signal is used in conjunction with synchronous demodulators, as shown in Figure 6.15. The transmitter subcarrier is generated continuously. A sample of this subcarrier of 9 ± 1 cycles is taken once per line and superimposed on the back porch of each line sync pulse, as shown in Figure 6.25. This subcarrier sample is called the **colour burst.**

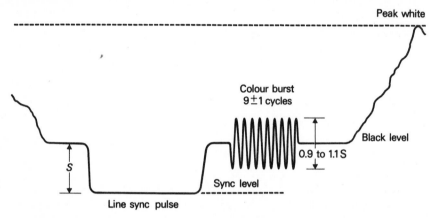

Figure 6.25 NTSC video signal during line blanking period. In the case of PAL the colour burst is specified as 10 ± 1 cycles

At the receiver the colour burst is extracted from the composite video signal and used to control the local subcarrier oscillators. Figure 6.26 shows a block diagram of the type of circuit that can be used. The line sync pulse separator produces an output during the trailing edge of the sync pulse which triggers the burst gate control pulse generator. This produces a pulse which opens the burst gate. The burst gate is effectively a pulse controlled switch. It only allows the video signal through to the local subcarrier generator during the colour burst. The 9 ± 1 cycles of colour burst are sufficient to lock the local subcarrier oscillator to the transmitter subcarrier both in phase and in frequency for the duration of a line.

The U axis has been chosen as phase reference, that is in the receiver, phase is measured relative to the phase of the subcarrier modulated by the $E'_B - E'_Y$ chrominance signal. The colour burst could be in the same phase, but this would not be the optimum choice. The burst occurs just before the start of each line scan and may be visible in the form of a vertical bar on the left of the picture.

240

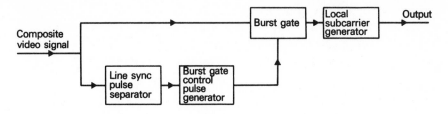

Figure 6.26 Use of burst gate to ensure that the video signal can only reach the local subcarrier generator when the colour burst signal is present

If the colour burst has amplitude A, and phase θ with respect to the U axis it will give rise to signals $E'_U = A \cos \theta$ and $E'_V = A \sin \theta$. The luminance, and hence visibility, of the colour burst can be worked out, as a multiple of A, for all possible values of θ. The smallest multiple is obtained when θ is about 180°. This is the phase chosen for the colour burst signal to ensure minimum visibility. The colour burst phase is shown in Figure 6.22. It is along the negative direction of the U-axis. A phase shifting circuit is used in receivers in order to provide reference subcarrier signals at the appropriate phase for the chrominance synchronous demodulators.

6.8 THE CHOICE OF SUBCARRIER FREQUENCY FOR THE NTSC SYSTEM

Imagine a scene of uniform luminance and chrominance, such as a uniformly lit, single colour, card filling the field of view of the camera. During the unblanked part of the line scans, the luminance signal will have a constant value; so will the chrominance signals, and the modulated colour subcarrier will have a constant amplitude and phase. The composite video signal will consist of a sinusoidal carrier superimposed on the d.c. luminance level as shown in Figure 6.27. Furthermore, this pattern will be repeated each line.

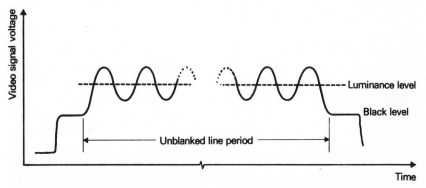

Figure 6.27 Video signal for a scene of uniform luminance and chrominance

241

The luminance channel of a colour receiver contains a notch filter to reject the subcarrier frequency. Some monochrome receivers do not have such a filter and the sinusoidal variation of the video signal produces a series of bright and dark dots along each line. These form a regular pattern which is more or less visible, depending on the relation between the line frequency, f_l, and the chrominance subcarrier frequency, f_{sc}. Various relations give minimum visibility. The one chosen in the NTSC system is

$$f_{sc} = \tfrac{1}{2}(2n + 1)f \qquad\qquad (6.13)$$

where n is an integer.

There is, therefore, an odd number of half cycles of carrier frequency in one line, and since the total number of lines in a complete picture is odd, it takes two pictures (four fields) for the complete pattern of dots to repeat itself. This can be seen by studying the simple raster of Figure 6.28 which

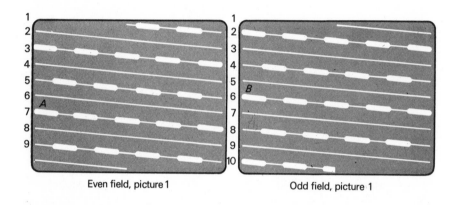

Even field, picture 1   Odd field, picture 1

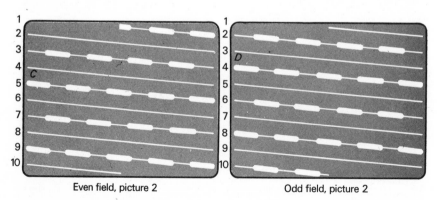

Even field, picture 2   Odd field, picture 2

Figure 6.28 Dot pattern due to chrominance subcarrier. A sequence of dots such as *A. B, C, D* is seen as a single dot moving vertically upwards

242

consists of 9 complete lines (4½ per field) and for which $n = 4$, giving $f_{sc} = 4½ f_l$. In each consecutive field the dots are staggered both vertically and horizontally with respect to the previous field. There is therefore, on average, no static pattern of bright and dark patches. There is, however, a stroboscopic effect which causes the spots to appear to move up and down the screen at the rate of one line per field. This dot 'crawl' can be visible in monochrome pictures, but only in regions corresponding to extended areas of uniform colour. Rapid changes of colour cause shifts in the phase of the modulated subcarrier which prevent regular, and hence discernable, patterns from being formed.

The higher the subcarrier frequency is, the finer the dot pattern will be, and hence the less visible. This frequency is therefore chosen to be as high as possible. It cannot be the highest frequency of the video band, because allowance must be made for the bandwidth of the chrominance signal. We will now consider how this affects the choice of subcarrier frequency.

The bandwidth of the various components of the American 525-line NTSC system are shown in Figure 6.29. The spacing between the vision and the sound carriers in the monochrome system is 4.5 MHz and this has to be preserved in the colour system for compatibility.

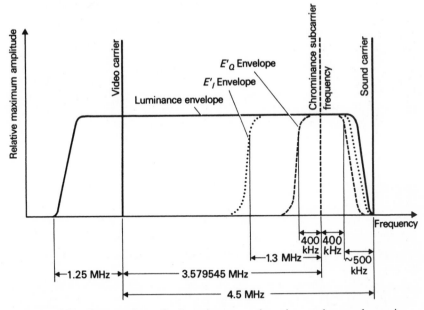

Figure 6.29 Frequencies of chrominance subcarrier and sound carrier relative to video carrier, and bandwidth of luminance and chrominance signals for the NTSC system

The E'_Q signal has a specified 2 dB bandwidth of 0·4 MHz. It is double sideband modulated onto the (suppressed) chrominance carrier. The wider band E'_I signal has a specified 2 dB bandwidth of 1·3 MHz, but in order to allow the chrominance subcarrier frequency to be as high as possible, vestigial side band modulation is used, with the upper side band attenuated. Note that this attenuation only starts beyond the upper edge of the E'_Q band. This is to avoid quadrature crosstalk which was discussed in Section 6.5.1. The asymmetry of the E'_I modulated chrominance signal band causes quadrature distortion on demodulation. The quadrature distortion contributes signals which would not be distinguishable from E'_Q signals, if they were produced in the same frequency band, and would therefore give rise to crosstalk. In order to avoid this, the symmetry of the modulated E'_I signal is preserved over the whole of that part of the frequency band which it shares with the E'_Q modulated signal.

As shown in Figure 6.29, the vestige of the E'_I band starts about 0·5 MHz above the chrominance subcarrier and drops to a negligible level over a further 0·5 MHz. The spacing between the chrominance subcarrier and the sound carrier must therefore be at least 1 MHz. The maximum spacing between the video carrier and the chrominance subcarrier is therefore about 3·5 MHz.

For minimum visibility on monochrome receivers, the chrominance subcarrier must be chosen to be exactly related to the line frequency according to Equation (6.13). This can be done conveniently by using frequency dividing circuits at the picture source in order to generate the line frequency directly from the subcarrier frequency. But the line and field frequency also have an exact relation. For a 60 Hz field frequency, and a 525-line picture, the line frequency is $\frac{1}{2} \times 60 \times 525 = 15\,750$ Hz and, again, it is useful to generate the field frequency by direct division of the line frequency. The supply mains frequency at the picture source could be multiplied up to generate the line and chrominance subcarrier frequencies, but the mains frequency can vary from place to place and it is better to use a stable electronic frequency standard so as to make possible interworking between distant studios.

The choice of subcarrier frequency f_{sc} is therefore subject to the following constraints

(a) f_{sc} must be as high as possible, that is about 3·5 MHz.

(b) $f_{sc} = \frac{1}{2}(2n + 1)f_l$, where f_l is about 15 750 Hz, and where $2n + 1$ is a convenient number for frequency division purposes, that is a number that has simple factors.

(c) The field frequency $= f_l/(525 \times 30)$ must be very close to 60 Hz so that the difference does not affect monochrome receivers.

These constraints are met by choosing a subcarrier frequency of 3·579 545 MHz and a value of 455 for $2n + 1$, with factors 5, 7 and 13. This gives a line frequency of $3\,579\,545 \times 2/455 = 15\,734\cdot264$ Hz, which is sufficiently close to 15 750 Hz; and a field frequency of $3\,579\,545\,(2/455)\,(2/525) = 59\cdot94$ Hz.

The subcarrier frequency, and the field frequency, are specified to about 1 part in $3\cdot5 \times 10^5$. The $59\cdot94$ Hz field frequency is well within the tolerance of the 60 Hz supply mains frequency used for monochrome receivers and is therefore compatible with it.

6.9 NTSC TRANSMITTERS AND RECEIVERS

We have now looked at the choice of the principal parameters for the NTSC system and are in a position to see how they lead to the structure of receivers and transmitters.

First, some of the parameters for the American 525-line system will be summarised.

The chrominance signals E'_I and E'_Q are formed from the gamma corrected camera outputs E'_R, E'_B and E'_G and, by Equations (6.11) and (6.12):

$$E'_Q = E'_V \sin 33° + E'_U \cos 33°$$
$$E'_I = E'_V \cos 33° - E'_U \sin 33°$$

where

$$E'_U = 0\cdot877 (E'_R - E'_Y)$$
$$E'_V = 0\cdot493 (E'_B - E'_Y)$$

and E'_Y, the luminance signal, is given by Equation (6.7)

$$E'_Y = 0\cdot30 E'_R + 0\cdot59 E'_G + 0\cdot11 E'_B$$

The 2 dB bandwidths of the E'_Q and E'_I signals are $0\cdot4$ and $1\cdot3$ MHz respectively. These signals are quadrature modulated on to a $3\cdot579 545$ MHz chrominance subcarrier. Balanced modulators are used so that the subcarrier is suppressed. A colour burst signal at the subcarrier frequency is sent during the back porch of each line blanking period. It is used to generate a reference subcarrier in the receiver.

The luminance and the modulated chrominance signals are combined, together with sync pulses, to form the composite video signal. This is then amplitude modulated onto an r.f. carrier in the v.h.f. or u.h.f. bands. A vestigial sideband filter is used to attenuate the lower sideband. The sound is frequency modulated onto a carrier $4\cdot5$ MHz above the video carrier.

The phase reference for the chrominance signals is taken as the U axis of Figure 6.22. In terms of this reference, the phases of the subcarrier onto which the E'_Q and E'_I signals are modulated are $33°$, and $90° + 33° = 123°$, respectively. The colour burst phase is $180°$.

The line and field frequencies are obtained by electronic division of the chrominance subcarrier frequency.

Figure 6.30 is a block diagram of a transmitter which can provide signals with the above properties. The function of most of the transmitter elements should be clear from the specification of the signals which they are intended to generate.

The low pass filters, used to band-limit the chrominance signals, introduce delays. Narrow band filters introduce more delay than wide band ones. The E'_Q signal is therefore delayed with respect to the E'_I signal and both are delayed with respect to the E'_Y luminance signal. The E'_Y and E'_I signals are given extra delays in order to make their total delays equal to that of the E'_Q signal.

The chrominance subcarrier generator produces two outputs differing in phase. The phases are shown referred to the U axis in Figure 6.22. One signal is taken to the colour burst gate which is an electronic switch operated by line sync pulses in such a way as to only allow the signal through during the back porch period. This ensures that the 9 ± 1 cycles of colour burst are incorporated into the right part of the composite video signal, as shown in Figure 6.25.

The second subcarrier signal, phase shifted by 33° with reference to the U axis, is fed to the E'_Q balanced modulator and, after being phase shifted by a further 90°, to the E'_I balanced modulator. This part of the circuit is equivalent to Figure 6.14, with the band pass filters incorporated into the modulators.

Figure 6.31 shows the basic elements of an NTSC colour receiver. The r.f. and i.f. stages, the envelope detector, sound receiver, and video amplifier serve the same functions as the corresponding elements of the monochrome receiver shown in Figure 5.33. The E'_I and E'_Q synchronous demodulators include low pass filters. As in the transmitter, extra delays are introduced into the paths of the E'_Y and E'_I signals in order to equalise the E'_Y, E'_I and E'_Q delays.

The subcarrier notch filter in the luminance channel is a narrow band-stop filter tuned to the chrominance subcarrier frequency. It is used to minimise interference by the chrominance signal on the luminance.

The burst blanking circuit is driven by a suitably timed pulse from the line scan generator. It acts as a switch, and is used to stop the burst signal from reaching the chrominance circuit during the back porch period. Without it, a vertical coloured stripe would be visible on the left-hand side of the picture.

The colour killer circuit is used to prevent monochrome transmissions, with components at the chrominance subcarrier frequency, from producing spurious colours, instead of a normal black and white picture. It cuts out the chrominance circuits if there are no colour bursts present in the received signals. This ensures that the colour difference signals are zero and that the same signal E'_Y is applied to all three electron guns.

246

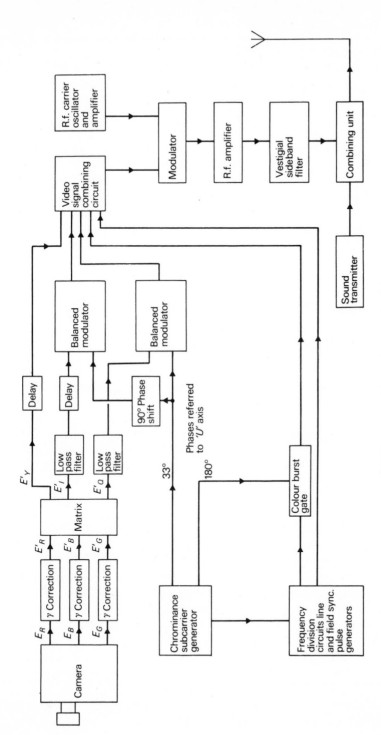

Figure 6.30 **Block diagram of an NTSC transmitter**

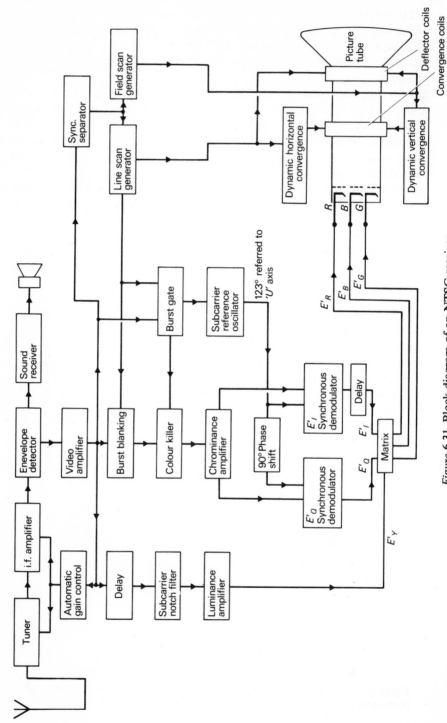

Figure 6.31 Block diagram of an NTSC receiver

The reference subcarriers for the synchronous demodulators must be in phase with the subcarrier used to modulate the E'_I and E'_Q signals at the transmitter. The E'_I subcarrier is at 123° to the U axis (Figure 6.22). This is the phase chosen for the E'_I demodulator. Some receivers use subcarrier phases corresponding to the U and V axes. One demodulated signal is then directly proportional to the colour difference signal $(E'_R - E'_Y)$ and the other to $(E'_B - E'_Y)$. This allows a simpler matrix circuit to be used than with I-Q demodulation. However, if this is done, both demodulators deal with a combination of E'_I and E'_Q signals and the E'_I signals have to be limited to the same bandwidth as the E'_Q signals, thereby sacrificing some of the chrominance resolution available in the NTSC system. Receivers of this type are sometimes called narrowband chrominance receivers, whereas receivers of the type shown in Figure 6.31 are known as broadband.

6.9.1 Differential phase distortion in the NTSC system

Figure 6.32 is a chrominance phasor diagram for a signal with components E'_I and E'_Q whose resultant OP makes an angle α with the U axis. The hue of the colour depends on the angle α. It can be seen, from Figure 6.22, that the hue of OP is magenta. The angle α is determined in the receiver from the phase difference between the local subcarrier generated from the colour burst and the modulated E'_I and E'_Q signals. The phase reference for the colour burst is the negative U axis.

The phasor OP represents a chrominance signal consisting of a sinusoid of amplitude OP, at the subcarrier frequency, differing in phase by $(180° - \alpha)$ from the colour burst. Phase shifts in the transmission channel, or in the receiver itself, may alter this phase difference so that it is no longer the same as it was at the transmitter. This is known as **differential phase distortion.** If differential phase distortion changes the phase difference by an angle β, say, the signal in Figure 6.32 which, at the transmitter, corresponds to OP, will

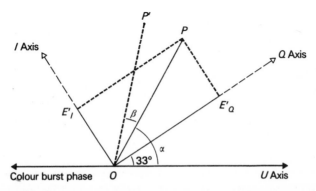

Figure 6.32 Transmitted chrominance signal OP received as signal OP' because differential phase distortion introduces phase error β.

be interpreted, at the receiver, as signal OP' which has the wrong hue. It appears too red for the positive value of β chosen in the diagram.

The amplitude of the colour burst (Figure 6.25) is kept constant to within 10 per cent. Not only does the amplitude of the modulated chrominance signal vary, but so does its level, because it rides on the luminance signal (see, for instance, Figure 6.20). Non-linearities in receiver circuits will cause some phase change when the mean signal level changes, and hence some differential phase distortion between the colour burst and the modulated chrominance signal. This is known as **level-dependent phase distortion.**

User tests indicate that differential phase distortion becomes objectionable when it exceeds about 10°. This cannot be considered as a precise figure, because there is a large variation in reported results.

6.10 THE PAL SYSTEM

Only the most significant differences between the PAL and NTSC systems and the main reasons for these differences will be discussed in this section. The PAL system was mainly developed to cope with the effects of differential phase distortion.

Both systems use $(E'_R - E'_Y)$ and $(E'_B - E'_Y)$ chrominance signals and a separate E'_Y luminance signal. Also they both use quadrature modulation for the chrominance signals, but the details of their modulation schemes are different.

Firstly, PAL systems normally operate on 625-line systems, which have more available bandwidth than the 525-line system used with NTSC. This allows more bandwidth for the chrominance signals, and it is not necessary to use a narrower band E'_Q signal and a wider band E'_I signal. It is more convenient to use equal-bandwidth signals and to operate directly with the E'_U and E'_V signals defined as in Equation (6.11).

$$E'_U = 0\cdot877(E'_R - E'_Y), \quad E'_V = 0\cdot493(E'_B - E'_Y)$$

Secondly, the E'_U signal is modulated on to a (suppressed) subcarrier whose phase corresponds to the positive direction of the U axis. The E'_V signal is similarly modulated on a quadrature subcarrier, but the phase of this subcarrier is reversed on alternate lines. This is shown in Figure 6.33. If the E'_V signal for the nth line is modulated onto a subcarrier whose phase is represented by the positive direction of the V axis, then the phase of the subcarrier used for the $(n + 1)$th line is represented by the negative direction of the V axis. The lines whose E'_V modulation is along the positive direction of the V axis will be called N lines. (N standing for NTSC, because the chrominance signals for these lines are coded in the same way as NTSC signals.) The others will be called P lines (P standing for PAL, because the coding for these lines is peculiar to the PAL system).

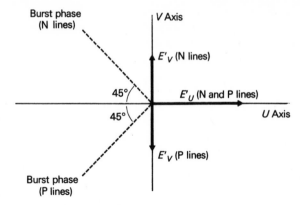

Figure 6.33 Chrominance subcarrier and burst phases for N and P lines in the PAL system

Thirdly, the phase of the colour burst is changed on alternate lines, from 135° with reference to the U axis phase on N lines, to 225° on P lines. It is called a **swinging burst** because of this alternation. As shown in Figure 6.33, the burst phase alternates by $\pm 45°$ about the phase of the colour burst in the NTSC system.

6.10.1 Operation of a PAL chrominance decoder
The basic elements of a PAL chrominance decoder circuit are shown in Figure 6.34.

The 64 μs (one line period) delay is usually obtained by ultrasonic propagation through a piece of glass fitted with input and output piezoelectric transducers. The delay allows the signals for consecutive lines, that is one N and one P line, to be combined in adding and subtracting circuits. The outputs of these circuits are fed to two synchronous demodulators. One has a reference subcarrier with the U phase so that its output is proportional to the U phase component of the modulated signal at its input. The other demodulator has a reference subcarrier which is electronically switched between the V and the $-V$ phases on alternate lines. The way this can be done will be described later, but first the phasor diagrams of Figure 6.35 will be used to explain how addition and subtraction of consecutive chrominance signals for consecutive lines can reduce the effects of differential phase distortion.

Let the transmitted modulated chrominance signal be represented by the phasor OP with components E'_{UT} and E'_{VT}, the suffix T standing for transmitted, along the U and V axes for an N line (Figure 6.35(a)). Its components for a P line are also E'_{UT} and E'_{VT} but with the phase of the V component reversed. (Figure 6.35(b)).

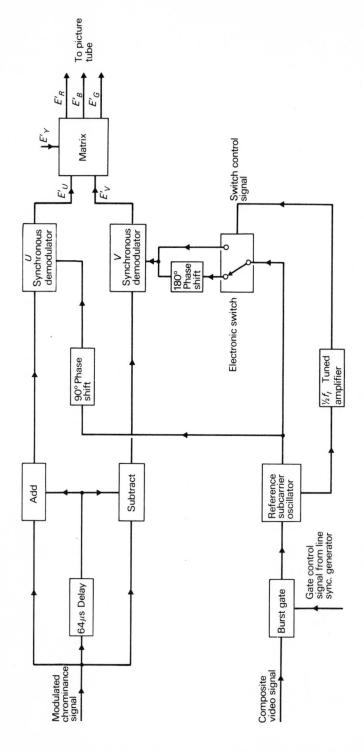

Figure 6.34 Block diagram of a chrominance decoder circuit in a PAL receiver.

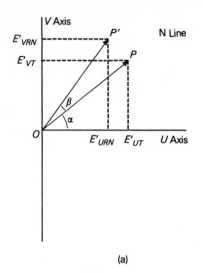
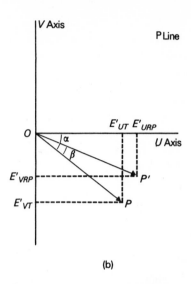

(a)   (b)

Figure 6.35 Effect of differential phase error β on *N* and *P* lines in the PAL system (a) for N lines; (b) for P lines. The transmitted signal *OP* is received as *OP'* because of differential phase error β.

Now, imagine that, because of differential phase distortion during transmission, the received signal (phasor *OP'*) is shifted by angle β in the positive direction for both *N* and *P* lines. Phase distortion does not change the length of the phasors so $OP = OP'$. We shall assume that the chrominance of the signal source does not change significantly from line to line, which means that both *OP* and α have the same value on consecutive lines. From Figure 6.35(a), the received *U* component for the *N* line, E'_{URN} , where the suffixes *R* and *N* stand for received and *N* line, is given by

$$E'_{URN} = OP' \cos(\alpha + \beta)$$
$$= OP \cos\alpha \cos\beta - OP \sin\alpha \sin\beta$$
$$= E'_{UT} \cos\beta - E'_{VT} \sin\beta$$

Similarly, the *V* component of the received *N* line, E'_{VRN} is given by

$$E'_{VRN} = E'_{VT} \cos\beta + E'_{UT} \sin\beta$$

The *U* and *V* components of the *P* line can be obtained in the same way using Figure 6.35(b), remembering that we now have (α − β) instead of (α + β). They are

$$E'_{URP} = E'_{UT} \cos\beta + E'_{VT} \sin\beta$$
$$E'_{VRP} = E'_{VT} \cos\beta - E'_{UT} \sin\beta$$

Note that the V axis intercepts are treated as positive quantities in Figure 6.35(b) because they do not represent ordinary coordinates but phasor amplitudes which are always taken as positive. The output of the U demodulator is the U component of the sum of the N and P line received signals, that is $E'_{URN} + E'_{URP} = 2 E'_{UT} \cos \beta$, irrespective of whether it is the N or the P line which has been delayed.

The difference between the V components of the two signals is $\pm (E'_{VRN} - E'_{VRP}) = \pm 2 E'_{VT} \cos \beta$, depending on which is subtracted from which, so that there is a sign change on consecutive lines. However, this is taken care of by switching the phase of the reference subcarrier to the V synchronous demodulator so that it is $0°$ for one line and $180°$ for the next, in the correct sequence to ensure that the modulator output is $2 E'_{VT} \cos \beta$ for all the line sequences.

The transmitted signals are E'_{UT} and E'_{VT} and the signals at the demodulator outputs are $2 E'_{UT} \cos \beta$ and $2 E'_{VT} \cos \beta$. The factor 2 does not matter; it can be accounted for by adjusting the gain of the chrominance amplifiers. The phase distortion has been removed since the modulator outputs correspond to a chrominance phase angle of $\tan^{-1} {}'e'_{VT} / E'_{UT}$, which is the same as the transmitted phase. However, the amplitude of the demodulated signals has been reduced in the proportion $\cos \beta : 1$. Thus, the phase error has been transferred into an amplitude error, which causes a change in saturation. This is much less objectionable than the change of hue which it replaces.

The reason for having a swinging colour burst is to provide a means of identifying N and P lines, so that the electronic switch controlling the reference phase for the V demodulator in Figure 6.34 can be correctly operated. The reference subcarrier oscillator produces an output at the chrominance subcarrier frequency, but it is controlled by the colour burst, via the burst gate. The colour burst goes through a complete phase swing cycle every two lines, that is at half the line frequency f_l. The reference subcarrier oscillator output also contains a component at this frequency which is in the right phase to operate the switch at the start of each line, and is used for this purpose after narrow band amplification, as shown in Figure 6.34.

With the exception of the chrominance circuits, PAL and NTSC receivers have essentially the same basic structure, shown in Figure 6.31.

A variety of alternative circuits are used for PAL receivers. One type, known as simple PAL, uses no delay line in order to reduce costs. The signals for consecutive pairs of lines cannot be summed in the receiver, but the effects of hue errors are reduced by the integrating action of the human eye. The reduction is only partial and, as far as users are concerned, this type of receiver is substantially less acceptable than a PAL receiver using delay line averaging.

6.10.2 The choice of chrominance subcarrier frequency for the PAL system

The subcarrier dot pattern is most noticeable in areas of uniform colour, corresponding to d.c. values of the chrominance signal. We will therefore only consider d.c. values. The E'_U chrominance signal is modulated onto a subcarrier whose phase does not alter from line to line. The E'_U modulated signal for areas of uniform colour is therefore a sinusoid at the subcarrier frequency f_{sc} and a suitable choice of subcarrier frequency for minimum visibility on monochrome receivers is $\frac{1}{2}(2n + 1) f_l$, as in the NTSC case.

However, the reference phase of E'_V is inverted every line. The modulated E'_V signal behaves effectively as two sinusoids differing in frequency by $f_l/2$, and the best choice of frequency for the one is the worst for the other.

A compromise choice is to use a quarter line offset for the dot pattern. The selected frequency is

$$f_{sc} = (284 - \tfrac{1}{4})f_l + 25 \text{ Hz} \tag{6.14}$$

The 25 Hz is an offset at the picture frequency. It produces a field interlace which helps to reduce subcarrier visibility.

A value of 4·43361875 MHz is chosen for f_{sc} in the 625 line, 50-field systems. This is specified to ± 1 Hz in the United Kingdom and ± 5 Hz in some other systems. With this value of f_{sc}, Equation (6.14) gives 15·625 kHz for the line frequency.

6.11 THE SECAM SYSTEM

The two colour difference signals are transmitted *sequentially* in the SECAM system. This is the most significant difference between SECAM and the PAL and NTSC systems in which the two signals are transmitted *simultaneously*.

In the SECAM system only one colour difference signal is transmitted during each line period. The red and blue difference signals are transmitted alternately. The complete chrominance signal is made up at the receiver by combining the signal which is being received at the time with the one which was transmitted during the previous line and stored in a memory device. The name SECAM (SEquentiel Couleur A Mémoire) comes from the sequential transmission of chrominance information and the use of a memory.

The memory device is a delay circuit, or delay line, as it is usually called. It is similar to the one used in PAL receivers and delays signals by one line period ($64 \, \mu s$) thus, enabling signals from consecutive transmitted lines to be combined, as shown in Figure 6.36. The switch is operated by pulses derived from the line sync generator.

The transmitter chrominance signals are chosen to be

$$D'_R = -1\cdot9\,(E'_R - E'_Y)$$
$$D'_B = 1\cdot5\,(E'_B - E'_Y)$$

255

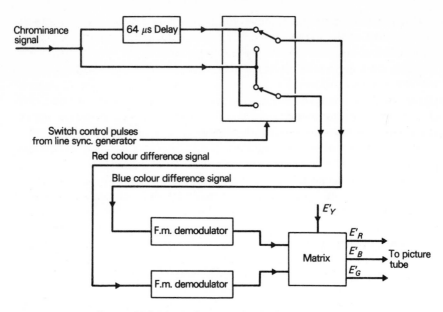

Figure 6.36 Block diagram of a SECAM decoder.

The luminance r.f. and i.f. circuits are similar to those of the other two systems and the same luminance E'_Y signal is used.

The chrominance signals are both limited to a 6 dB bandwidth of 1·5 MHz. Frequency modulation is used. The D'_R signal frequency modulates a sub-carrier of (4·402 25 ± 0·002) MHz during one line, and the D'_B frequency modulates a subcarrier of (4·2500 ± 0·002) MHz during the next.

At the receiver, the operation of the switch (Figure 6.36) ensures that one f.m. demodulator always has the modulated D'_R signal at its input (delayed one line, directly the next) and the other demodulator always has the modulated D'_B signal at its input.

Figure 6.36 is very much simplified. Pre-emphasis is used at the transmitter for both the unmodulated and the modulated chrominance signals in order to improve the signal-to-noise ratio. De-emphasis is used in the receiver but this is not shown in the figure, because we are only concerned here with the fundamental differences between SECAM and the other two systems. These differences are the transmission of only one chrominance signal at a time and the storing of the other signal in the receiver, so that both are simultaneously available at the matrix input.

Problems of differential phase distortion are avoided by transmitting the two frequency modulated chrominance signals separately. This avoids having to use quadrature amplitude modulation. The f.m. demodulators used to recover the video chrominance signals are much less complex than the synchronous demodulators of the NTSC and PAL receivers.

When one chrominance signal is being transmitted, the other is not, so that only half of the chrominance information available from the camera is transmitted at any time. Either the red or the blue colour difference signals reaching the picture tube are the same for any two consecutive lines. If it is the chrominance information corresponding to the red colour difference signal which is changed in going from line n to line $(n + 1)$, then the chrominance information corresponding to the blue colour difference signal is the same for both these lines. It changes for line $(n + 2)$ which has the same red colour difference information as line $(n + 1)$. The vertical chrominance resolution of the picture is therefore half that of NTSC or PAL pictures with the same number of lines, but this is no real disadvantage.

The reason for this goes back to the discussion of video bandwidth and resolution in Section 5.6 of Chapter 5. If the number of lines, and hence the vertical resolution, of a picture are fixed, then changes in bandwidth can only affect the horizontal resolution. The chrominance bandwidth is less than the luminance bandwidth in all television systems, because we perceive less colour detail than brightness detail. In the NTSC and PAL systems, all the reduction in chrominance resolution takes place in the horizontal direction because the chrominance and luminance components of the picture are both built up from the same number of lines. In Section 5.6, we saw that a 625-line system with a luminance bandwidth of 5·5 MHz has roughly equal vertical and horizontal *luminance* resolution. Thus in the PAL system, with a chrominance bandwidth of 1·3 MHz, the vertical *chrominance* resolution is greater than the horizontal chrominance resolution by a factor of $5·5/1·3 \simeq 4$. This factor is reduced to about 2 in a 625-SECAM system with similar bandwidths. But even so, the effective limit to the perception of colour detail is still set by the horizontal resolution, which is approximately the same as in the other two systems.

6.12 CONCLUSION

The three main colour television systems and their several variants are used in national broadcasting networks throughout the world and it is tempting to try and compare them. However, such a comparison is certain to be misleading, and it is more instructive to consider why this should be so.

There is no immediately obvious and striking difference between the quality of the pictures and the degree of compatibility obtained from the three systems. They all have defects, the importance of which can be judged, at least in principle, by user tests. In practice, it is difficult to draw conclusions. For instance, the relative susceptibility of NTSC and SECAM receivers to interference by sinusoidal signals is found to depend on the frequency of these signals or, again, the relative visibility of echo effects in the two systems depends on the length of the transmission delays causing the echoes.

User tests have been applied successfully to individual systems in order to decide on specific requirements, but the results of tests designed to compare systems are rarely found to be conclusive. Comparative figures should be treated with caution, and one does not have to read many published comparisons to realise that the author's allegiances to one system may have a significant effect on the results quoted.

Systems can be compared in terms of cost, particularly receiver costs which are likely to have a direct effect on the number of viewers. Only short term comparisons are likely to be reliable because it is always possible that new production techniques may change the situation. For instance, the cost of the delay lines used in the PAL and SECAM system would have been prohibitive at the time the NTSC system was being developed.

Broader economic factors are very relevant, though hard to assess. Existing investments in research, development and plant, and the possibility of export markets can be important considerations.

Finally, national and international politics can play a significant part. Pressure groups representing the interests of companies and administrations are likely to form, and some of these may have a significant effect. It is also conceivable that the decision to adopt, or even not to adopt, the same system as that used in another country may be the determining factor.

Broader considerations of this type always affect the usefulness and viability of telecommunication systems and system designers cannot afford to ignore them. This book has been mainly limited to technological and short term cost considerations, but it should not end without stressing the importance of human, economic and political factors in the field of telecommunication systems.

Appendix

Table A.1 *Picture and signal standards for the principal monochrome television systems*

	British 405-line system	American 525-line system	British 625-line system	European CCIR-625 line system	French 819-line system
Number of lines per picture	405	525	625	625	819
Field frequency (Hz)	50	60	50	50	50
Picture frequency (Hz)	25	30	25	25	25
Aspect ratio	4/3	4/3	4/3	4/3	4/3
Line frequency (Hz)	10125	15750	15625	15625	20475
Line period (μs)	98·8	63·5	64	64	48·84
Active field factor a_1 (see Section 5.6)	0·931	0·923	0·922	0·922	0·919
Active line factor a_2 (see Section 5.6)	0·814	0·826	0·812	0·812	0·805
Number of active picture lines	377	485	575	575	753
Video bandwidth (MHz)	3	4·2	5·5	5	10
Number of picture elements which can be resolved along a line drawn vertically down the screen (Kell factor = 0·7)	264	340	402	402	527
Number of picture elements which can be resolved along a horizontal line	483	422	572	520	786
Transmitted gamma (approximate)	2 to 2·5	2·2	2·2	2·2	1·7
Sense of modulation	positive	negative	negative	negative	positive
Black level as % of peak carrier	35	75	77	75	25
Blanking level as % of peak carrier	30	75	77	75	25
Peak white level as % of peak carrier	100	15	20	10	100
Bandwidth of r.f. transmission channel (MHz)	5	6	8	7	14
Attenuated (vestigial) sideband	upper	lower	lower	lower	upper
Bandwidth of Nyquist flank region on either side of video carrier (MHz)	0·75	0·75	1·25	1·25	2
Position of sound carrier with respect to vision carrier (MHz)	−3	+4·5	+6	+5·5	−11·15
Sound modulation	a.m.	f.m.	f.m.	f.m.	a.m.
Sound carrier deviation (kHz)		±25	±50	±50	
Sound pre-emphasis (μs)		75	50	50	

References and Bibliography

General background and telecommunication principles

1 *Communication*, A Scientific American Book (Reprint of *Scientific American* September 1972). San Francisco: W. H. Freeman and Co.
2 Pierce, J. R., 1962. *Symbols, Signals and Noise*. London: Hutchinson.
3 Brown, J. and E. V. D. Glazier, 1974. *Telecommunications*, 2nd edn. London: Chapman and Hall.
4 Carlson, A. B., 1968. *Communication Systems*. New York: McGraw-Hill.
5 Schwartz, M., 1970. *Information Transmission, Modulation and Noise*. New York: McGraw-Hill.
6 King, R. W. P., H. R. Mimmo and A. H. Wing, 1965. *Transmission Lines, Antennas and Wave Guides*. New York: Dover.
7 Atkinson, J., 1970. *Telephony*, Vol. 1. London: Pitman.
8 Wakling, P. J., 1972. *Pulse Code Modulation*. London: Mills and Boon.
9 Cattermole, K. W., 1969. *Principles of Pulse Code Modulation*. London: Iliffe.
10 Bennett, W. R. and J. R. Davey, 1965. *Data Transmission*. New York: McGraw-Hill.

General aspects of telecommunication systems

11 C.C.I.T.T., 1972. *Proceedings of the Fifth Plenary Assembly of the C.C.I.T.T.* (The Green Books) Geneva: International Telecommunication Union.
12 Davies, D. W. and D. L. A. Barber, 1973. *Communication Networks for Computers*. London: John Wiley.
13 Hills, M. T. and B. G. Evans, 1973. *Telecommunications Systems Design*, Vol. 1. London: George Allen & Unwin.
14 Jolley, E. H., 1968. *Introduction to Telephony and Telegraphy*. London: Pitman.
15 Martin, J., 1970. *Teleprocessing Network Organisation*. New Jersey: Prentice-Hall.
16 Martin, J., 1969. *Telecommunications and the computer*. New Jersey: Prentice-Hall.
17 Panter, P. F., 1972. *Communication Systems Design: Line-of-sight and Troposcatter Systems*. New York: McGraw-Hill.
18 Squires, T. L., 1970. *Telecommunications Pocket Book*. London: Newnes-Butterworth.

The performance of switched systems

19 Franklin, R. H., 1961. World-wide telephone transmission, *Institution of Post Office Electrical Engineers Paper No. 222*, 3rd October.
20 Kusunoki, S. *et al.*, 1974. Analogue transmission performance on the switched telecommunications network, *Japan Telecommunications Review* 16, part 2, 96-104.
21 Mellors, W. J. G., 1967. Telephone performance measurement, *G.E.C. Telecommunications* 36, 23-35.
22 Richards, D. L., 1968. Transmission performance of telephone networks containing p.c.m. links, *Proc.I.E.E.* 115, part 9, 1245-58.
23 Richards, D. L., 1973. *Telecommunication by Speech*. London: Butterworths.
24 Zeidler, G., 1973. Coin operated telephones: an analysis and appraisal of existing types leading to a new design concept, *Electrical Communication* 48, part 3, 260-8.

Terminals in switched systems

25 Roberton, J. S. P., 1956. The rocking-armature receiver, *Post Office Electrical Engineers Journal* **49**, 40-6.

26 Smith, S. F., 1974. *Telephony and Telegraphy A*. London: Oxford University Press.

27 Spanton, J. C. and P. L. Connellan, 1969. Modems for the Datel 200 service, *Post Office Electrical Engineers Journal* **62**, 1-10.

28 Spencer, H. J. C., 1955. Some principles of anti-sidetone telephone circuits, *Post Office Electrical Engineers Journal* **48**, 208-11.

Exchanges and Signalling

29 Atkinson, J., 1972. *Telephony*, Vol. 2. London: Pitman.

30 Bini, A., 1970. International telephone switching centres, *Electrical Communication* **45**, part 1, 4-12.

31 G.E.C., 1972. *G.E.C. Telecommunications* **39** (whole issue on s.p.c.).

32 Habara, K. *et al.*, 1971. Remote control electronic switching systems, *Review of the Electrical Communication Laboratory of Japan* **19**, part 3, 211-16.

33 Hamer, M. P. R., 1974. Reliability modelling considerations for a real-time control system, *Digest of papers of the I.E.E.E. International Symposium on Fault Tolerant Computing* 1974, 22-7.

34 Hills, M. T. *Principles of Telecommunications Switching Systems*. To be published.

35 Hobbs, M., 1974. *Modern Communications Switching Systems*. U.S.A.: Tab Books.

36 Horsfield, B. R., 1971. Fast signalling in the U.K. telephone network, *Post Office Electrical Engineers Journal*, **63**, part 4, 242-52.

37 Oden, H., 1970. Push-button selection: a characteristic of modern switching systems, *Electrical Communication* **45**, part 1, 66-71.

38 Electrical Communication Laboratory of Japan, 1974, *Review of the Electrical Communication Laboratory of Japan* **22**, parts 9 and 10 (whole double issue on s.p.c.).

39 Welch, S., 1964. Signalling systems for dialling over transoceanic telephone cables, *Institution of Post Office Electrical Engineers Paper No. 225*, 28th April.

Planning, traffic and economics

40 Andrews, F. T. and R. W. Hatch, 1971. National telephone network transmission planning in the American Telephone and Telegraph Company, *Trans. I.E.E.E.* **COM-19**, 302-15.

41 C.C.I.T.T., 1964. *National Telephone Networks for the Automatic Service (handbook)*. Geneva: International Telecommunication Union.

42 C.C.I.T.T., 1968. *Local Telephone Networks (handbook)*. Geneva: International Telecommunication Union.

43 Munday, S., 1967. New international switching and transmission plan recommended by the C.C.I.T.T. for public telephony, *Proc.I.E.E.* **114**, part 5, 619-27.

44 Syski, R., 1960. *Introduction to Congestion Theory in Telephone Systems*. London: Oliver and Boyd.

Transmission systems

45 Bell Telephone Laboratories, 1970. *Transmission Systems for Communications*. U.S.A.: Western Electric.

46 Bell Telephone Laboratories, 1969. The L4 coaxial system, *Bell System Technical Journal* **48**, part 4 (whole issue).

47 Boag, J. F. and J. B. Sewter, 1971. The design and planning of the main transmission network, *Post Office Electrical Engineers Journal* **64,** part 1, 16-21.
48 Halliwell, B. J., 1974. *Advanced Communication Systems.* London: Newnes-Butterworths.
49 Jones, D. G., 1964. The Post Office network of radio-relay stations. *Post Office Electrical Engineers Journal* **57,** part 3, 147-55.
50 Karbowiak, A. E., 1965. *Trunk Waveguide Communications.* London: Chapman and Hall.
51 I.E.E.E., 1970. Millimetric wave propagation, *I.E.E.E. transactions on antennas and propagation* **AP-18,** part 4 (whole issue).
52 Powell, J., 1975. International integrated civil communication systems, *I.E.E. Electronics and Power* 6th March, 249-52.
53 Swain, E. C., 1965. Local lines—past, present, and future, *Institution of Post Office Electrical Engineers Paper No. 226,* 8th November.
54 Taylor, F., 1971. Intelsat—the international telecommunications satellite, *I.E.E. Electronics and Power* January, 8-13.
55 Turner, D. and T. B. M. Neill, 1958. The principles of negative-impedance convertors and the development of a negative-impedance 2-wire repeater, *Post Office Electrical Engineers Journal* **51,** 206-11.

Textbooks on television
56 Hutson, G. H., 1966. *Television Receiver Theory,* Part 1. London: Edward Arnold. (monochrome only)
57 Wharton, W. and D. Howorth, 1967. *Principles of Television Reception.* London: Pitman. (monochrome and colour)
58 Fink, D. J., 1957. *Television Engineering Handbook.* New York: McGraw-Hill. (monochrome and NTSC colour)
59 Carnt, P. S. and G. B. Townsend, 1961. *Colour Television,* Vol. 1, Theory and Practice. London: Iliffe.
60 Carnt, P. S. and G. B. Townsend, 1969. *Colour Television,* Vol. 2, PAL, SECAM and Other Systems. London: Iliffe.
61 Herrick, C. N., 1973. *Colour Television.* Reston, Virginia: Reston Publishing Co.
62 Hutson, G. H., 1971. *Colour Television Theory.* London: McGraw-Hill.
63 Osborne, B. W., 1968. *Colour Television Reception and Decoding Techniques.* London: MacLaren and Sons.
64 Patchett, G. N., 1974. *Colour Television.* London: Norman Price.
65 Reed, C. R. G., 1969. *Principles of Colour Television Systems.* London: Pitman.
66 Townsend, B., 1970. *PAL Colour Television.* London: Cambridge University Press.

Television general topics, user tests, specifications
67 Bingley, F. J., 1953, 1954. Colorimetry in colour television, *Proc I.R.E.* Part I **41,** 838-51; part II **42,** 48-51; part III **42,** 51-7.
68 Hall, R. C. and J. Forrest, 1971. Colour planning for television studios, *Proc. Inst. Rad. and Electron Eng. Australia* **32,** part 5, 167-74.
69 Issue on the NTSC colour system 1954. *Proc. I.R.E.* **42,** part 1, 1-357 (a whole issue devoted to the development of the NTSC system).
70 Luxenberg, L. R. and R. L. Kuehn (eds), 1968. *Display Systems Engineering.* New York: McGraw-Hill.
71 Pawley, E., 1972. *BBC Engineering 1922-1972.* London: BBC publications.
72 *Report of the Television Advisory Committee* 1972. London: H.M.S.O.
73 Allnatt, J. W. and R. D. Prosser, 1966. Subjective quality of colour television pictures impaired by random noise, *Proc. I.E.E.* **113,** part 4, 551-7.

74 DeCola, R., R. E. Shelby and K. McIlwayn, 1954. NTSC color field test, *Proc. I.R.E.* **42,** part 1, 20-43.

75 Weaver, L. E., 1959. Subjective impairment of television pictures, *Electronic and Radio Engineer* **36,** 170-9.

76 Pearson, D. E., 1975. *Transmission and Display of Pictorial Information.* London: Pentech Press.

77 NTSC signal specification 1954. *Proc. I.R.E.* **42,** part 1, 17-19.

78 *Specification of Television Standards for 625-line System, I. Transmissions* 1971. London: IBA and BBC joint publication.

79 Technical performance targets for a PAL colour television broadcasting chain 1969. *The Radio and Electronic Engineer* **38,** part 4, 201-16. (Based on papers prepared by the PAL working party of the British Radio Equipment Manufacturers' Association.)

Television transmission

80 Bartlett, H. F., 1973. Design trends in television transmitters, *Proc. Inst. Rad. and Electron. Eng. Australia* **34,** part 9, 390-400.

81 Heightman, 1971. Gamma in television, *Sound and Vision Broadcasting* **12,** part 2, 26-30.

82 Lewis, N. W., 1954. Waveform response of television links, *Proc. I.E.E.* **101,** part 3, 258-70.

83 MacDiarmid, I. F., 1959. Waveform distortion in television links, *Post Office Electrical Engineers Journal* **52,** part 2, 108-14; **52,** part 3, 188-95.

84 Paddock, F. J., 1970. The relationship between individual links and chain distortions in a television network, *I.E.R.E. Conference Proc.* No. 18, 127-41.

85 Redmond, J., 1974. Cable and radio programme links in the BBC, *BBC Engineering* **97,** 5-15.

86 Steele, H., G. McKenzie and R. Vivian, 1970. The impact of automation on television transmission, *The Royal Television Society Journal* **13,** part 6, 131-8.

87 Teesdale, G. A. R. and E. M. Hickin, 1972. Microwave links for television outside broadcasts, *The Royal Television Society Journal* **14,** 107-12.

88 Weaver, L. E., 1971. *Television Measurement Techniques.* London: Peter Peregrinus.

Camera tubes and colour cameras

89 Bailey, P. C., 1970. New lead-oxide tubes, *Sound and Vision Broadcasting* **11,** part 2, 19-21.

90 Lord, A. V., 1971. Advances in colour television cameras, *Electronics and Power* **17,** 337-42.

91 Turk, W. E. and E. D. Hendry, 1973. The development of the camera television tube, *Broadcasting Technology.* London: I.E.E. Publications, 21-41.

92 Capers, J. D. and P. W. Loose, 1970. The mark VIII camera channel-design of the video circuits, *Sound and Vision Broadcasting* **11,** part 2, 24-30.

93 Cosgrove, M. and K. Schat, 1974. The digital control of a colour camera and its systems implications, *The S.E.R.T. Journal* **8,** 75-7.

94 Cuomo, A. C., 1970. A digitally controlled colour television camera, *J. of the S.M.P.T.E.* **79,** 1003-8.

95 James, I. J. P., D. G. Perkins, P. J. Pyke, E. W. Taylor, D. E. Kent and I. A. Fairbairn, 1970. The E.M.I. four-tube colour television camera, *Radio and Electronic Engineer* **39,** part 5, 249-70.

96 Parker-Smith, N. N., 1972. Design of a modem automatic colour television camera, *G.E.C. Journal of Science and Technology* **39,** part 3, 98-106.

Index